7

D1797084

£12
(Twelve Pounds)
NM

PROGRESS IN HPLC
Volume 1

PROGRESS IN HPLC
Volume 1

Gel permeation and ion-exchange
chromatography of proteins and peptides

Editors
Hasan Parvez, Yoshio Kato, and Simone Parvez

Utrecht, The Netherlands
1985

VNU Science Press BV
P.O. Box 2073
3500 GB Utrecht
The Netherlands

© 1985 VNU Science Press BV

First published 1985

ISBN 90–6764–048–4

Printed in Great Britain

Preface

In recent years high-performance liquid chromatography (HPLC) has undergone remarkable growth and has now developed into a powerful technique for separation and characterization of a variety of substances. This technique is especially useful in dealing with hydrophilic substances, including many complex molecules of biological importance such as peptides, proteins, and enzymes.

The editors of this volume have been able to assemble many acknowledged experts whose efforts have led to continued improvements in the analytical applications of HPLC to peptides and proteins. Comprehensive accounts of their research experience with different aspects of HPLC as applied to peptides and proteins are described in several chapters of this book which may be regarded as a collection of critical reviews of the current use of HPLC in a very specialized area of research.

I am confident that the book will be found useful as a valuable reference work for researchers and serious students interested in any of the specific topics covered and that the many efforts involved will be justified.

December 1984

B. C. DAS
Institut de Chimie des Substances Naturelles
CNRS, Gif-sur-Yvette, France

This book is dedicated to
Sydney Udenfriend, teacher and friend,
and
Toshihara Nagatsu

CONTENTS

Preface   v

Introduction   ix

Application of TSKgel SW in nucleic acid separation
Y. KATO, H. PARVEZ, and S. PARVEZ   1

Membrane-bound hormone receptor purification and characterization
by HPLC
R. G. L. SHORR, M. W. STROHSACKER, R. REBAR,
R. J. LEFKOWITZ, M. G. CARON, and S. T. CROOKE   9

Determination of protein molecular weight by gel permeation
chromatography equipped with low-angle laser light scattering photometer
T. TAKAGI   27

Gel permeation of proteins by high-performance gel chromatography
in denaturing solvents
K. KONISHI   43

Multiple forms of cytochrome P-450 separable by high-performance
liquid chromatography
Y. FUNAE, A. N. KOTAKE, and K. YAMAMOTO   59

Gel permeation studies of glycoprotein by high-performance
liquid chromatography
Y. SHIMOHIGASHI and H.-C. CHEN   83

Gel permeation of human serum lipoprotein by HPLC
I. HARA and M. OKAZAKI   95

Heterogeneity of renin: application of HPLC in studies on renin and
renin binding protein
H. IWAO, S. KIM, Y. FUNAE, N. NAKAMURA, F. IKEMOTO,
and K. YAMAMOTO   105

Enzyme purification by high-performance ion-exchange liquid
chromatography
F. B. RUDOLPH, B. F. COOPER, and J. GREENHUT   133

Application of HPLC for analysis and purification of catecholamine-
synthesizing enzymes
K. KOJIMA, S. PARVEZ, H. PARVEZ, Y. KATO, and T. NAGATSU   149

High-performance liquid chromatography for enkephalins and
enkephalin-containing peptides
K. KOJIMA and T. NAGATSU   157

Technical aspects of biochemical high-performance column liquid
chromatography
J. SJÖDAHL, K. J. DILLEY, R. ERIKSSON, J. PELLERIN,
S. H. PARVEZ, and L. ARLINGER   179

Author index   219

Subject index   221

Introduction

Separation of proteins by low-pressure chromatography using Sephadex-type supports has greatly facilitated biomedical research during the past two decades. However, the length of time required to achieve a separation by such methods often leads to a great loss of biological activity. Some proteins manifest hydrolysis even at 4°C, separation interval is even longer, greater is the disintegration of a specific molecule of interest. The introduction of TSK gels by Toyo Soda Laboratories revolutionized this particular field of analytical chemistry. Their introduction in biomedical research received a great impulse due to conservation of biological activity at the maximal level. It is interesting to mention that most of SW-type stationary phases work with physiological buffers at pH 7; this avoids biological activity modulation. Ion exchange is another solution to protein separation but salt loading and variability of pH can lead to denaturation of protein molecules. The main achievement in gel permeation at high performance has been the stability of the matrix which can easily resist to 100 bars. This allows us to have a complete profile of protein separation in the molecular weight range of 5000 to 1 000 000 in 15–60 min.

This volume provides a collection of papers from international specialists on all aspects of protein separation. The book is a true reflection of an international cooperation in biomedical and physical research. The preface written by Professor B. C. Das, Director, Protein Laboratory CNRS, France, analyses all the highlights of this volume. The novelty of this contribution is also associated with additions of papers on absolute molecular determination by laser scattering devices in gel permeation as well as application of this methodology to membrane-bound receptor proteins. The book describes in detail modern methodology to separate nucleic acids, enzymes and a wide variety of biologically active proteins such as renin. The editors express their sincere gratitude to all contributors for making this book a success. We hope that it will serve as a valuable reference in protein analysis.

H. PARVEZ
Y. KATO
S. PARVEZ

Progress in HPLC, Vol. 1, pp. 1—7
Parvez *et al.* (Eds)
© 1985 VNU Science Press

Application of TSKgel SW in nucleic acid separation

YOSHIO KATO,[1]* HASAN PARVEZ,[2] and SIMONE PARVEZ[2]

[1] Central Research Laboratory, Toyo Soda Mfg. Co., Ltd, Tonda, Shinnanyo,
Yamaguchi, Japan and [2] Unité de Pharmacologie, Université de Paris XI,
Centre d'Orsay-Bat. 440, 91405 Orsay Cedex, France

INTRODUCTION

Aqueous gel permeation HPLC was introduced several years ago and has been a widely accepted technique in protein separation. In many separations, TSKgel SW was used successfully, however, its application in the field of nucleic acid has been very limited so far (Wehr and Abbott, 1979; Uchiyama *et al.*, 1981; Graeve *et al.*, 1982). Accordingly, we investigated the TSKgel SW in respect of its separation range, and its resolution and operational variables in gel permeation HPLC of RNAs and double-stranded DNA fragments (Kato *et al.*, 1983a, b, 1984). The results are described here.

EXPERIMENTAL

Total *E. coli* RNA was purchased from Miles Laboratories (Elkhart, Ind., USA). This sample contains four components: 4 S tRNA (mol. wt 25 000), 5 S rRNA (39 000), 16 S rRNA (560 000), and 23 S rRNA (1 100 000). *E. coli* tRNA, type XXI was purchased from Sigma (St Louis, Mo., USA). Double-stranded DNA fragments were prepared by cleaving plasmid DNA pBR322 with restriction endonuclease HaeIII or BstNI. The preparation method has been described elsewhere (Kato *et al.*, 1984). HaeIII-cleaved pBR322 DNA contains 22 fragments of 7, 11, 18, 21, 51, 57, 64, 80, 89, 104, 123, 124, 184, 192, 213, 234, 267, 434, 458, 504, 540, and 587 base pairs. BstNI-cleaved pBR322 DNA contains six fragments of 13, 121, 383, 928, 1060, and 1857 base pairs. The molecular weights of the fragments were determined by multiplying their base pair numbers by 650.

Gel permeation HPLC was carried out at 25°C with a high-speed liquid chromatograph Model HLC-803D, equipped with a variable-wavelength u.v. detector (Toyo Soda, Tokyo, Japan). The u.v. detector was operated at 260 nm. Refractive

---

* To whom correspondence should be addressed.

index detector was used in the study of sample loading. Column systems consisting of two G2000SW columns, two G3000SW columns, or two G4000SW columns (Toyo Soda) were used. Each column was 60 × 0.75 cm i.d. The eluent was usually 0.1 M phosphate buffer (pH 7.0) containing 0.1 M sodium chloride and 1 mM EDTA. Tris-HCl buffer (pH 7.5) containing sodium chloride and EDTA was also used in some cases. The flow rate was 1 ml/min except in the study of its effect. The injection volume and sample concentration were usually 0.1 ml and 0.01–0.1%.

RESULTS AND DISCUSSION

Separation range

Chromatograms of total *E. coli* RNA and HaeIII-cleaved pBR322 DNA are shown in Figs 1 and 2. All peaks were fractioned and assigned by polyacrylamide gel electrophoresis. Figures 3 and 4 show the molecular weight calibration curves for RNA and DNA fragment obtained from Figs 1 and 2. For peaks of DNA fragments containing more than one component, the average molecular weights were adopted. These calibration curves demonstrate that RNAs and DNA fragments were separated mainly on the basis of their molecular weights. The exclusion limits of columns estimated from the calibration curves are summarized in Table 1, indicating that RNAs of molecular weight up to 1 500 000 and double-stranded DNA fragments of molecular weight up to 300 000 can be separated on TSKgel SW.

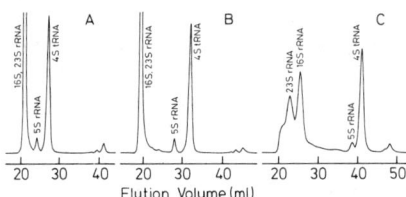

Figure 1. Chromatograms of total *E. coli* RNA obtained on column systems of two G2000SW columns (A), two G3000SW columns (B), and two G4000SW columns (C) in 0.1 M phosphate buffer (pH 7.0) containing 0.1 M sodium chloride and 1 mM EDTA at a flow rate of 1 ml/min.

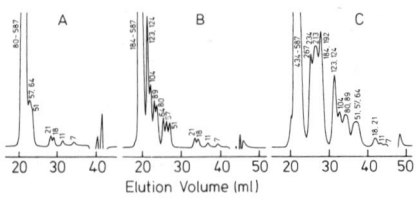

Figure 2. Chromatograms of HaeIII-cleaved pBR322 DNA obtained on column systems of two G2000SW columns (A), two G3000SW columns (B), and two G4000SW columns (C) in 0.1 M phosphate buffer (pH 7.0) containing 0.1 M sodium chloride and 1 mM EDTA at a flow rate of 1 ml/min. The numerals above the peaks indicate the base pairs of DNA fragments contained in the peaks.

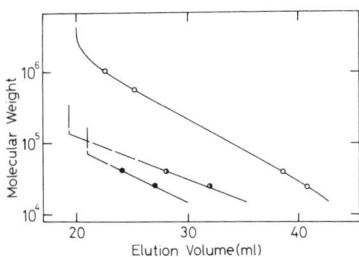

Figure 3. Molecular weight calibration curves of two G2000SW (●), two G3000SW (◑), and two G4000SW (○) column systems for RNAs in 0.1 M phosphate buffer (pH 7.0) containing 0.1 M sodium chloride and 1 mM EDTA.

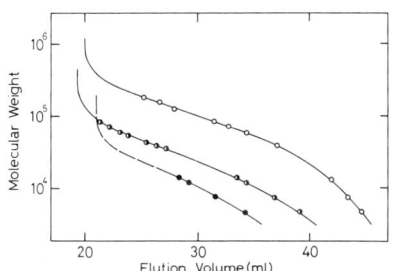

Figure 4. Molecular weight calibration curves of two G2000SW (●), two G3000SW (◑), and two G4000SW (○) column systems for double-stranded DNA fragments in 0.1 M phosphate buffer (pH 7.0) containing 0.1 M sodium chloride and 1 mM EDTA.

Table 1. Exclusion limit of G2000SW, G3000SW, and G4000SW for RNA and double-stranded DNA fragment in 0.1 M phosphate buffer (pH 7.0) containing 0.1 M sodium chloride and 1 mM EDTA

| | Exclusion limit in molecular weight | |
| | RNA | Double-stranded |
Column		DNA fragment
G2000SW	70 000	50 000
G3000SW	150 000	100 000
G4000SW	1 500 000	300 000

Separation efficiency

Figures 1 and 2 demonstrate that RNAs and DNA fragments can be separated with high efficiency on TSKgel SW. G2000SW and G3000SW can easily separate components differing by more than 10% in molecular weight in less than 1 h. Consequently, it may be possible to adopt gel permeation HPLC on TSKgel SW as an alternative to gel electrophoresis, which has been the most popular technique to separate nucleic acids. It should be particularly useful for preparative purposes because the separated components in samples can be recovered easily and yet almost quantitatively.

The separation of a sample depends on the column used, as can be seen from Figs 1 and 2. Therefore, it is very important to select the best column depending on the molecular weights of the samples to be separated. Tables 2 and 3 summarize the best columns, established by visually comparing the chromatograms in Figs 1 and 2.

Table 2. Best columns for the separation of RNSs

Molecular weight range	Best column
<60 000	G2000SW or G3000SW
60 000–120 000	G3000SW
120 000–1 500 000	G4000SW

Table 3. Best columns for the separation of double-stranded DNA fragments

Molecular weight range	Best column
<40 000	G2000SW or G3000SW
40 000–80 000	G3000SW
80 000–300 000	G4000SW

Effect of eluent ionic strength

Figure 5 shows the effect of eluent ionic strength on the elution volumes obtained on G4000SW. A similar effect was observed on G2000SW and G3000SW as well. Elution of both RNAs and DNA fragments was delayed by increasing the eluent

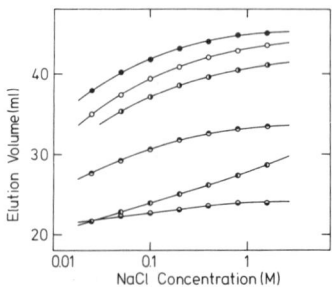

Figure 5. Dependence of elution volume on eluent ionic strength obtained with total *E. coli* RNA [4 S tRNA (○), 5 S rRNA (◑), 16 S rRNA (◐)], and BstNI-cleaved pBR322 DNA [fragments of 13 (●), 121 (◒), and 383 (◓) base pairs] on two G4000SW column system in 0.01 M Tris-HCl buffer (pH 7.5) containing 0.025–1.6 M sodium chloride and 1 mM EDTA at a flow rate of 1 ml/min.

ionic strength. Elution volumes greatly varied in the low ionic strength region, but at high ionic strength the elution volumes seemed to become constant. Furthermore, the elution volumes of small molecules were more markedly affected than those of large molecules. The peak widths broadened very slightly with increasing eluent ionic strength. Accordingly, an eluent ionic strength of 0.3–0.5 may be appropriate in general. When the eluent of low ionic strength is used, the exclusion limit of the column becomes considerably low, as shown in Fig. 6.

The main source of variation of elution volume with eluent ionic strength is probably the repulsive ionic interaction between samples and supports, because both nucleic acids and TSKgel SW are negatively charged. (TSKgel SW is based on silica and contains some residual silanol groups on its surface.) However, other sources such as adsorption on the support may also be responsible in the case of 16 S and 23 S rRNAs because their elution volumes increased regularly with eluent ionic strength, even in the high ionic strength region where ionic interactions would be expected to diminish.

Effect of flow rate
Figure 7 shows chromatograms of total *E. coli* RNA obtained on G4000SW at different flow rates. The separation was significantly improved by decreasing flow rate. Figure 8 shows the dependence of height equivalent to a theoretical plate (HETP) on flow rate. The HETP decreased with decreasing flow rate throughout the range investigated. Particularly in the case of high molecular weight sample (16 S rRNA), the HETP greatly depended on flow rate and reached a minimum at

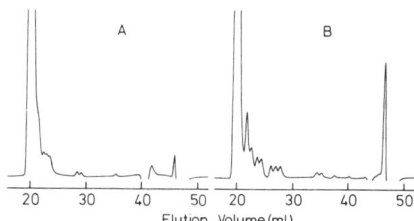

Figure 6. Chromatograms of HaeIII-cleaved pBR322 DNA obtained on two G4000SW column system in 0.01 M Tris-HCl buffers (pH 7.5) containing 1 mM EDTA and 0.05 M (A) or 0.4 M (B) sodium chloride at a flow rate of 1 ml/min.

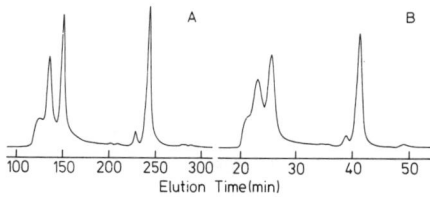

Figure 7. Chromatograms of total *E. coli* RNA obtained on two G4000SW column system in 0.1 M phosphate buffer (pH 7.0) containing 0.1 M sodium chloride and 1 mM EDTA at flow rates of 0.17 ml/min (A) and 1 ml/min (B).

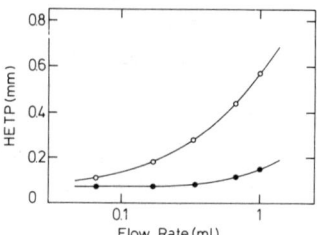

Figure 8. Dependence of HETP on the flow rate obtained with total *E. coli* RNA [4 S tRNA (●), 16 S rRNA (○)] on two G4000SW column system in 0.1 M phosphate buffer (pH 7.0) containing 0.1 M sodium chloride and 1 mM EDTA.

flow rates below 0.1 ml/min. Flow rates of 0.3–0.5 ml/min seem to be a good compromise when separation time and resolution are taken into consideration.

Effect of sample loading
Figure 9 shows the dependence of HETP on sample loading obtained for tRNA on G4000SW. HETP was almost constant up to 5 mg and increased with further increasing sample loading. In the separation of total *E. coli* RNA, almost the same chromatograms were obtained with sample loadings of 0.06 and 0.5 mg. Therefore, it is likely that nucleic acids can be separated without any decrease in resolution up to a few milligrams.

CONCLUSIONS

Gel permeation HPLC on TSKgel SW was found to be useful for the separation of RNAs and double-stranded DNA fragments. RNAs of molecular weight up to 1 500 000 and double-stranded DNA fragments of molecular weight up to 300 000 could be separated with fairly high resolution. Elution of RNAs and DNA fragments was significantly influenced by eluent ionic strength. Eluent ionic strength of 0.3–0.5 seemed to be appropriate in general. Resolution greatly increased

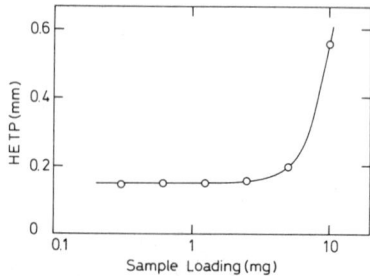

Figure 9. Dependence of HETP on the sample loading obtained with *E. coli* tRNA on two G4000SW column system in 0.05 M Tris-HCl buffer (pH 7.5) containing 0.2 M sodium chloride and 1 mM EDTA at a flow rate of 1 ml/min with a constant injection volume of 0.2 ml and varying sample concentration.

with decreasing flow rate, especially in the case of high molecular weight samples. RNAs could be separated without decrease in resolution up to a few miligrams.

REFERENCES

Graeve, L., Goemann, W., Földi, P., and Kruppa, J. (1982). Fractionation of biologically active messenger RNAs by HPLC gel filtration. *Biochem. Biophys. Res. Commun. 107*, 1559–1565.

Kato, Y., Sasaki, M., Hashimoto, T., Murotsu, T., Fukushige, S., and Matsubara, K. (1983a). Operational variables in high-performance gel filtration of DNA fragments and RNAs. *J. Chromatogr. 266*, 341–349.

Kato, Y., Hashimoto, T., Murotsu, T., Fukushige, S., and Matsubara, K. (1983b). Loading capacity in high-performance gel filtration of nucleic acid on TSKgel SW. *HRC & CC 6*, 626.

Kato, Y., Sasaki, M., Hashimoto, T., Murotsu, T., Fukushige, S., and Matsubara, K. (1984). A new packing for separation of DNA restriction fragments by high performance liquid chromatography. *J. Biochem. 95*, 83–86.

Uchiyama, S., Imamura, T., Nagai, S., and Konishi, K. (1981). Separation of low molecular weight RNA species by high-speed gel filtration. *J. Biochem. 90*, 643–648.

Wehr, C. T. and Abbott, S. R. (1979). High-speed steric exclusion chromatography of bio-polymers. *J. Chromatogr. 185*, 453–462.

Progress in HPLC, Vol. 1, pp. 9—26
Parvez *et al.* (Eds)
© 1985 VNU Science Press

Membrane-bound hormone receptor purification and characterization by HPLC

ROBERT G. L. SHORR,[1] MARK W. STROHSACKER,[1] RICHARD REBAR,[1]
ROBERT J. LEFKOWITZ,[2] MARC G. CARON,[2] and STANLEY T. CROOKE[1]

[1] Smith Kline & French Laboratories, Philadelphia, PA 19101, USA and [2] Howard Hughes Medical Institute, Duke University, Durham, NC 27710, USA

INTRODUCTION

A principle tenet in pharmacology is that neurotransmitters, drugs, and hormones interact with specific receptor sites (proteins, carbohydrates, DNA, lipids) to produce or inhibit a characteristic physiological response. In the case of synaptic transmission and many hormone responsive systems, these specific receptor sites are presumed to be integral membrane proteins. To date, only a few of these plasma membrane-bound neurotransmitter/hormone receptors have been purified to homogeneity. These include the acetylcholine receptor from a variety of sources such as electroplax (Moore *et al.*, 1979; Raftery *et al.*, 1980) and skeletal muscle (Shorr *et al.*, 1981a; Gotti *et al.*, 1982), the epidermal growth factor receptor (Carpenter, 1983) and the β-adrenergic receptors from avian (Shorr *et al.*, 1982a, b), amphibian (Shorr *et al.*, 1981b), and mammalian sources (Benovic *et al.*, 1982; Homcy *et al.*, 1983). β-Adrenergic receptors are part of a functional multi-protein complex which modulates the enzyme adenylate cyclase in response to agonists or hormone. The pharmacological identity of these receptors is based upon their specific interactions with the compounds isoproterenol (iso), epinephrine (epi), and norepinephrine (norepi). These ligands are agonists which stimulate the production of cyclic AMP by inducing a transient interaction between receptor and a guanine nucleotide binding protein (Stadel *et al.*, 1982). This protein appears to act as a shuttle between receptor and enzyme, regulating agonist promoted stimulation of cyclic AMP synthesis (Stadel *et al.*, 1981). β-Adrenergic receptor subtypes are delineated based upon the relative order of potency of these compounds in producing their effect, (i.e.) for β_1 iso > norepi $\geqslant$ epi and for β_2 iso > epi $\gg$ norepi.

Routinely, β_1- and β_2-adrenergic receptors are detected in plasma membrane preparations using a number of high-specific radioactivity receptor directed ligands. These include [^{125}I] cyanopindolol (Engel *et al.*, 1981), [^{125}I] hydroxybenzyl-pindolo (Levitski *et al.*, 1974), and [^{3}H] dihydroalprenolol (Lefkowitz *et al.*, 1974). These compounds are non-selective in their affinities for β_1 and β_2 receptors.

Non-specific binding is defined in the presence of excess concentrations of un-labeled agonist or antagonist.

Subsequent examination of the ligand binding properties of β-adrenergic receptors using a number of synthetic agonist and antagonist compounds has confirmed the discrimination of these plasma membrane-bound receptors into two subtypes; β_1 and β_2. Both proteins stimulate adenylate cyclase activity in response to agonist or hormone. Additionally there are α-adrenergic receptors which recognize these compounds with essentially the opposite rank order of potency of β receptors; i.e., norepi $>$ epi $\gg$ iso. Relative to adenylate cyclase, the α receptors function to inhibit the production of cyclic AMP, presumably through a separate guanine nucleotide binding protein (Rodbell, 1980). For both α and β receptors the agonist and antagonist ligands act stereospecifically with the $(-)$isomer more potent than the $(+)$isomer.

We are currently examining the structural and biochemical basis of the phar-macological distinctions between the β_1 and β_2 receptor subtypes. This has, of course, necessitated the development of new techniques for receptor purification and subsequent characterization after extraction from the membrane. It is largely from the perspective of these methodologies that this chapter is presented, against a background of those problems inherent in the purification of integral membrane proteins, i.e., hydrophobicity, the necessary presence of detergents, and those problems which are perhaps particular to the β_1- and β_2-adrenergic receptors from turkey and frog erythrocyte plasma membranes, respectively.

MATERIALS AND METHODS

Materials

Alprenolol hydrochloride was generously supplied by Hassle Pharmaceutical Co. (Sweden) and was coupled to Sepharose 4B–C1 (Pharmacia) as described in Caron et al. (1979). Digitonin was obtained from Gallard-Schlesinger (New York) and prepared as described in Shorr et al. (1981a, b). Ligands for receptor assay; [^{3}H]-dihydroalprenolol; [^{125}I] cyanopindolol as well as carrier-free Na[^{125}I] and [^{125}I]-paraazidobenzylcarazolol were from New England Nuclear. Premixed SDS poly-acrylamide gel electrophoresis standards (phosphorylase b, M_r = 94 000; bovine serum albumin, M_r = 67 000; ovalbumin, M_r = 43 000; carbonic anhydrase, M_r = 30 000; soybean trypsin inhibitor, M_r = 20 100; α-lactalbumin, M_r = 14 000) were obtained from Pharmacia and iodinated as in Shorr et al. (1981a, b). Protease inhibitors bacitracin, soybean trypsin inhibitor, benzamidine, and phenylmethyl-sulfonyl fluoride were obtained from Sigma Chemical Co. Electrophoresis reagents were obtained from Bio-Rad or Baker Chemical Co. High performance liquid chromatography was carried out using a Waters Associates (Bedford, MA) Model 6000A pump with U6K injector and Model 440 detector at 280 nM, or with a Beckman Model 344 liquid chromatograph system. Toyosoda TSK-3000SW and TSK-4000SW columns (8 $\times$ 600 mm) were obtained from Kratos (New Jersey) or Altex (CA). Protein I-125 and I-250 size exclusion columns were from Waters Associates.

Methods

Preparation of purified erythrocyte membranes. Purified frog erythrocyte membranes were prepared as described in Caron and Lefkowitz (1976) except that protease inhibitors (bacitracin, 100 μg/ml; soybean trypsin inhibitor, 10 mg/ml; benzamidine 10^{-4} M; and phenylmethylsulfonyl fluoride, 3 $\times$ 10^{-5} M) were included. Membranes were frozen in liquid nitrogen and stored without loss of receptor activity at $-90°C$ until use.

Turkey red blood cell membranes were also prepared in the presence of protease inhibitors as described in Shorr *et al.* (1982a, b). Briefly, whole turkey blood (1.2–2L) was obtained (Featherdown Farms, Apex, NC) and the red blood cells washed free of serum and leukocytes with buffered saline. Packed cells (600–800 ml) were then stirred for 90 min (25°C) in a precooled (4°C) Parr nitrogen bomb (Parr Instrument Co., IL) under 800 psi pressure. Lysis was achieved by releasing cells to atmospheric pressure. Nuclei were removed by centrifugation (4°C) at 500g for 20 min. Membranes were collected by centrifugation of the supernatant at 40 000g for 15 min and washed three times with 75 mM Tris-HCl pH 7.4, 25 mM $MgCl_2$, prior to suspension in 600–800 ml of the same buffer and freezing in liquid nitrogen.

Preparation of solubilized β-adrenergic receptor. Once detected in a suitable tissue, the first step in isolation of receptors is solubilization of the protein from plasma membrane preparations with retention of biological activity; in this case, radioligand binding with characteristic orders of affinities and stereoselectivity of the defining adrenergic agents described above.

To date, for the β-adrenergic receptor, the only method suitable for this purpose is membrane dissolution with the detergent digitonin. Other more common detergents such as Lubrol, Triton X-100, Tween, cholate, and deoxycholate although capable of solubilizing receptor results in a form that does not allow direct interaction with receptor directed ligands. Once solubilized from the membrane, neither $β_1$- nor $β_2$-adrenergic receptors are linked to adenylate cyclase.

Briefly, particulate receptors from frog or turkey red cell membranes are solubilized with 2% digitonin [100 mM NaCl, 10 mM Tris-HCl pH 7.2, 3 $\times$ 10^{-5} M phenylmethylsulfonyl fluoride, 2 $\times$ 10^{-4} M benzamidine, 100 μg/ml bacitracin, 10 μg/ml soybean trypsin inhibitor and, in some preparations, 2 $\times$ 10^{-4} M N-ethylmaleimide and 5 or 10 $\times$ 10^{-3} M bis(aminoethyl ether)-N,N,N^1,N-tetraacetic acid (EDTA)] , by resuspending the pellet obtained from 200–300 ml of membranes in 80 ml of the detergent buffer and Dounce homogenizing. This procedure is repeated twice, final volumes adjusted to 300 ml with detergent buffer and membranes stirred on ice for 45 min. Unsolubilized material is removed by centrifugation at 240 000g for 45 min using a Beckman 45 Ti rotor with receptor binding activity remaining in the supernatant. Alternatively residual particulate material can be removed by centrifugation at 40 000g for 30 min in a Beckman JA-221 centrifuge with JA10 rotor.

Receptor assays. Both soluble and particulate receptor assays for turkey and frog red cell preparations using [^{3}H] dihydroalprenolol (DHA; 31.4 Ci/mmol) are as in Shorr *et al.* (1981a, b, 1982a, b). For soluble assays free ligand is separated from bound ligand by Sephadex G-50 gel filtration. Binding competition curves generated for purified preparations using [^{125}I] cyanopindolol (CYP) are as in Shorr *et al.* (1982a, b) with data computer analyzed as described in De Lean *et al.* (1981). Labeling of receptor, purified or particulate, with the photoaffinity probe [^{125}I]-paraazidobenzylcarazolol (pABC) is as in Lavin *et al.* (1982). These antagonist ligands, [^{3}H]DHA, [^{125}I]CYP and [^{125}I]pABC, are nonselective in their binding properties to β_1 and β_2 receptor subtypes. Protein determinations for particulate and purified preparations are made as in Shorr *et al.* (1981a, b).

Sepharose-alprenolol chromatography. Detergent solubilized receptor activity (300 ml) is loaded (25°C) onto a 200 ml column at 200 ml/h and unbound protein washed through (0–4°C) with 1–2 column volumes of 0.05% digitonin, 100 mM NaCl, 10 mM Tris-HCl pH 7.2 with protease inhibitors. For β_2 (frog erythrocyte) preparations, receptor activity is eluted (25°C) with a front of (±)alprenolol (40 μM) and receptor in each fraction collected (15 ml) and activity measured by [^{3}H]DHA assay. Prior to assay, aliquots (200 μl) of each fraction are chromatographed on columns of Sephadex G-50 to remove excess eluting ligand. For β_1 (turkey erythrocyte) preparations, a linear gradient (300 ml) of 0–40 μM (±)alprenolol is used for elution from the affinity column and receptor activity measured as described above for β_2 preparations. In both cases, fractions containing receptor activity are pooled and concentrated a maximum of 30-fold by ultrafiltration (Amicon cell) with YM30 or PM30 membrane at 10–15 psi.

Gel permeation high performance liquid chromatography. Concentrated receptor from affinity column eluates are chromatographed on size exclusion columns using two 600 mm TSK-4000SW and one 600 mm TSK-3000SW columns tandem linked. Mobile phase containing detergent is prepared by first boiling a 2% digitonin suspension until clarification and allowing insoluble material to precipitate for at least 40 h at 4°C. The supernatant is filtered through 0.2 μm Millipore membranes and lyophilized. Solutions of 0.2% digitonin 100 μM Tris-SO$_4$ pH 7.2 (25°C) are prepared and refiltered through 0.2 μm Millipore membranes prior to use. Samples (2 ml) are manually injected and chromatographed at 1 ml/min, 300 μl fractions are collected 36 min after sample injection and aliquots assayed. Fractions containing receptor activity are pooled, concentrated and rechromatographed through a second HPLC run and assayed for activity and protein.

Alternatively for larger scale separations, affinity column preparations are pooled (150 ml), concentrated to 10 ml, and chromatographed as above utilizing two TSK-4000SW columns in tandem (25.4 × 600 mm^2). Flow rates are 5 ml/min with 0.1% digitonin 25 mM Tris-SO$_4$ pH 7.2, 10 mM EDTA as the mobile phase. Fractions of 2.5 or 5 ml are collected and receptor activity localized by assay as described above.

Gel permeation (octylglucoside/SDS) chromatography. To unfold receptor proteins, to achieve complete purity and to develop techniques for the separation of peptides by non-reverse phase methods, we examined the use of size exclusion columns equilibrated with SDS/octylglucoside buffers. Columns used were tandem linked TSK-4000, TSK-3000, and TSK-2000 (6 × 300 mm) in addition to a TSK-3000 guard column. Flow rates were 0.2 ml/min. Sample sizes were up to 500 μl and were detected at 220, 280 nM or by ^{125}I gamma counting. Optimal buffer conditions were 0.2% octylglucoside: 0.08% SDS, 25 mM Tris-HCl pH 6.8, 150 mM NaCl, 1 mM DTT.

Sodium dodecyl sulfate-polyacrylamide gel electrophoresis. SDS–PAGE of purified ^{125}I-labeled or [^{125}I] pABC labeled receptor was performed as in Shorr *et al.* (1981a, b, 1982a, b) using 10% or 12% homogeneous gels. Radiolabeling of purified preparations was as in Shorr *et al.* (1981a, b). Sample buffer consisted of 10% SDS, 12 mM Tris-HCl, pH 6.5, 10% glycerol, 5 mM dithiothreitol. Samples were denatured 30–60 min at 25°C in order to prevent aggregation. After electrophoresis, gels were dried using a Bio-Rad (Model 224) gel dryer and exposed (−90°C) with Kodak XAR-5 film with Lightning Plus (Dupont) intensifying screens. Films were developed manually using Kodak chemicals according to Kodak instructions.

Isoelectric focusing. Isoelectric focusing is an equilibrium electrophoretic technique by which proteins are separated according to their isoelectric point in a stable pH gradient. ^{125}I-labeled separated β receptor peptides from *Staphylococcus aureus* protease digests were focused using an LKB mutiphor apparatus according to O'Farrel (1975) except that gels contained 0.1% Nonidet 40. Ampholytes (pH 3.5–9.5) were obtained from LKB. Samples were focused at 30 W to a voltage of 1 800 V and the pH gradient measured with an LKB surface electrode. Precolored isoelectric focusing standards (Bethesda Research Labs) were utilized to confirm complete focusing. Gels were subsequently dried and autoradiographed using Kodak XAR-5 films as above.

RESULTS

Purification of β-adrenergic receptors from frog red blood cell plasma membranes
As described elsewhere, we have been successful in the development of an affinity support utilizing the compound alprenolol (Caron *et al.*, 1979). This potent receptor antagonist (K_D = 1 nM) is non-selective for β_1 and β_2 receptors and can be utilized for the affinity purification of both proteins. As shown in Table 1, use of such an affinity support in a repetitive fashion, coupled with ion exchange chromatography on DEAE Sepharose 6B–CL, results in the preparation of purified β_2-adrenergic receptor from frog erythrocyte membranes. This material on iodination with Na[^{125}I] and chloramine T and SDS–PAGE reveals a major band of protein centered at about 58 000 M_r, in addition to material migrating at lower molecular weight. Two-dimensional SDS–PAGE of such preparations, however, reveals that only the 58 000 band coincides with receptor activity (Shorr *et al.*, 1981a, b). We have also utilized the photoaffinity probe [^{125}I] parazidobenzyl-carazolol (Lavin *et al.*, 1982) which is also non-selective between β_1 and β_2 receptor subtypes to specifically label this receptor in both membrane and purified preparations (Shorr *et al.*, 1982b). In each case for particulate or purified preparations of frog erythrocyte plasma membrane receptors, a single specifically labeled band

Table 1. Summary of purification of the β-adrenergic receptor of frog erythrocytes by repetitive affinity chromatography

Step	Yield at each step (%)	Overall yield (%)	Specific activity (pmol/mg)	Purification (fold)
Crude frog erythrocyte membranes	100	100	0.15	1
Digitonin extract	70	70	1.0[a]	6.6
First alprenolol-gel pass	50–70	35–50	106[a]	706
Second alprenolol-gel pass	50	18–30	1470[a]	9800
DEAE-Sepharose	60–80	10–15	n.d.	n.d.
Final alprenolol-gel pass	30–50	4–8	8000–10 000	55 000

[a] Representative of the specific activities at these stages of the purification. This purification procedure has been carried out a number of times (n = 7) and has been found to yield 2–5 μg of measurable protein at the end of the four steps.

Loading of receptor activity on the affinity gels was performed batchwise. Gels were then washed in a column with 0.2% digitonin/100 mM NaCl/10 mM Tris-HCl, pH 7.4 and eluted in the same buffer containing 10 mM (+)isoproterenol and 50 μM DTT or 40 μM alprenolol for 5–6 h at 22–23°C. Eluate fractions containing receptor activity were pooled, lyophylized, resuspended in water (10–25 ml) and chromatographed on Sephadex G-50 to remove excess free alprenolol. Throughout the procedures shown, receptor activity was assayed by [^{3}H]DHA binding. In the case of alprenolol-gel eluates the sample was first desalted on Sephadex G-50. n.d. = not determined.

at 58 000 M_r is observed by SDS–PAGE. These results have been confirmed elsewhere (Rashidbaigi and Ruoho, 1982). Additionally, we have examined the molecular size of this protein in membrane and purified preparations by radiation inactivation and obtained nearly identical results to those shown here (Shorr *et al.*, 1984). Binding studies carried out utilizing purified preparations further clearly demonstrate the specific interaction of characteristic β-adrenergic ligands with receptor. This data is summarized in Table 2 where particulate, solubilized, and purified receptor affinities for adrenergic agents are displayed.

Although the procedures described for receptor isolation as outlined in Table 1 provide purified receptor protein, overall yields are extremely low (4–8%) and the process is prolonged and expensive.

Therefore, to develop a rapid high-yield procedure, we have adapted HPLC on steric exclusion columns for β receptor purification (Shorr *et al.*, 1982b). Figure 1 demonstrates the profile obtained when alprenolol eluates of the affinity column that are purified approximately 100-fold from frog erythrocyte membrane preparations (Table 1) are chromatographed using HPLC. An approximate 25-fold further purification of receptor binding activity is achieved (see Table 3). No significant purification of receptor is obtained on standard Sephadex/Sepharose filtration. As shown in Fig. 2, when fractions from the receptor binding peak are pooled, radioiodinated (Na[^{125}I]/chloramine T) and subjected to one or two additional HPLC runs, a single peak of radioactivity (corresponding to receptor binding activity which had been specifically prelabeled in the membranes prior

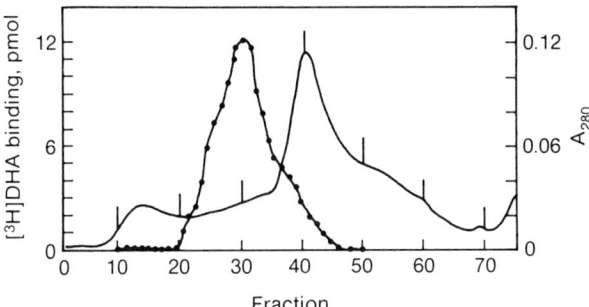

Figure 1. HPLC elution profile of partially purified β-adrenergic receptor preparation. The concentrated eluate of a Sepharose-alprenolol affinity gel was chromatographed on two I-250 and one I-125 tandem-linked columns (Waters Associates, Bedford, MA) (total volume, 36 ml). The flow rate was 1 ml/min and the mobile phase 0.1% digitonin, 100 mM Tris-SO_4 pH 7.2 at RT°. Fractions of 300 µl were collected and receptor activity located by [^{3}H]dihydroalprenolol binding assay. Protein was monitored by $A_{280 \text{ nm}}$. From Shorr *et al.* (1982a).

Table 2. Comparison of the affinities of purified, solubilized and membrane-bound frog erythrocyte β-adrenergic receptors for various β-adrenergic ligands

Agents	β-adrenergic receptor preparations		
	Membrane-bound [a] K_D (nM)	Solubilized [a] K_D (nM)	Purified K_D (nM)
(−)Alprenolol	3.4	4.1	1.25
(+)Alprenolol	150	220	56.0
(−)Isoproterenol	400	210	520
(+)Isoproterenol	183 000	68 000	79 700
(−)Epinephrine	4600	3700	2300
(−)Norepinephrine	49 000	41 000	56 000

[a] Taken from Caron *et al.* (1976).

K_Ds for the various drugs were calculated from competition curves for [^{3}H]DHA binding according to DeLean *et al.* (1981) assuming a K_D of 2.2 nM for [^{3}H]dihydroalprenolol.

to digitonin solubilization and HPLC) is obtained. Subsequent SDS–PAGE of the ^{125}I-labeled protein again reveals a single band at 58 000 M_r (Fig. 2 inset). Isoelectric focusing of this material also reveals a single peak of radioactivity at a pI of 5.8, identical to the pI of [^{125}I]hydroxybenzylpindolol prelabeled receptor or receptor purified by repetitive affinity chromatography (Shorr *et al.*, 1982a).

In Table 3 we give a summary of the purification of receptor which can be achieved by combining affinity chromatography and high performance size exclusion chromatography techniques. After complete purification (i.e., affinity chromatography and two HPLC runs) a specific activity ranging from 11 800 to 14 400 pmol of binding sites per mg protein is obtained with an average yield of 30% of the receptor binding sites in crude digitonin extracts. These biological

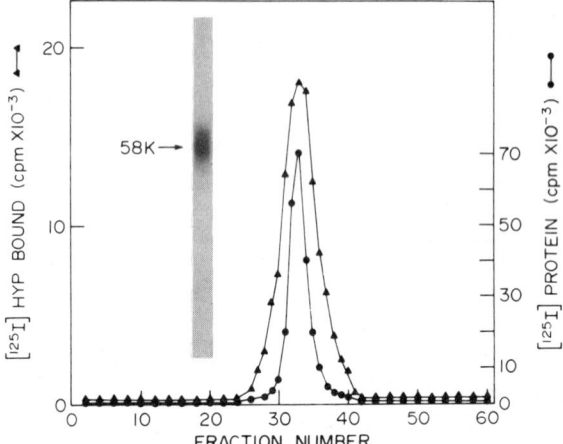

Figure 2. HPLC elution profile of radioiodinated, purified receptor (●) and [^{125}I]hydroxy-benzylpindolol labeled β-adrenergic receptor. (▲) A receptor preparation eluted from the Sepharose-alprenolol gel was chromatographed as described in Methods. The peak of receptor activity was pooled and iodinated by the chloramine T method (Shorr *et al.*, 1981a, b), desalted on Sephadex G-50 to remove unreacted Na[^{125}I] and chromatographed twice on two I-250 and one I-125 columns in tandem. Inset: pooled fractions of [^{125}I]β-adrenergic receptor preparation were electrophoresed on 12% homogeneous SDS-polyacrylamide slab-gel according to Methods. Samples were denatured by boiling 3 min with 10% SDS, 5% β mercaptoethanol 12 mM Tris-HCl pH 6.5 or by incubating 1 at 55°C with 10% SDS 2 mM DTT 12 mM Tris-HCl pH 6.5. (–▲–) A frog erythrocyte membrane preparation in which receptors were labeled with [^{125}I]hydroxybenzylpindolol, prior to solubilization with 1% digitonin, 100 mM NaCl, 10 mM Tris-HCl pH 7.2 and chromatography as described above. From Shorr *et al.* (1982a).

Table 3. Summary of purification of β-adrenergic receptor of frog erythrocytes by affinity and high performance liquid chromatography

Step	Activity (pmol)	Yield at each step (%)	Overall yield (%)	Specific activity, (−)[^{3}H]Dihydro-alprenol bound (pmol/mg)	Purification (fold) each step	overall
Detergent extract of frog erythrocyte membranes	396	100	100	1.9	1	1
Eluate of alprenolol affinity gel	245	62	62	136	72	72
First HPLC	206	84	52	3416	25	1800
Second HPLC	122	59	31	11 800	3.5	6300

Typically, membranes from 200–300 ml of frog erythrocytes were solubilized with digitonin and affinity chromatography and HPLC steps performed as described under Methods. Protease inhibitors at concentrations stated above were included up to the first HPLC step. The experiment shown is representative of five such experiments. The overall purification from crude frog erythrocyte membranes (0.15 pmol/mg) is >80 000 fold.

specific activities do, in fact, approach those predicted from radiation target analysis studies where an antagonist binding site size was determined to be 54 000 M_r in the frog red cell plasma membrane system (Shorr *et al.*, 1984). The purification folds obtained by these methods represent a 6500-fold purification from initial detergent extracts and a greater than 80 000-fold purification from membrane preparations. Finally, examination of these purified preparations which contain only the 58 000 M_r polypeptide by competition experiments with characterization ligands for [125 I] cyanopindolol binding reveals typical β_2-adrenergic specificity as exemplified in Table 2. Thus, comparison of the procedures described in Table 3 and Table 1 demonstrate that the purification of receptor from frog erythrocyte membranes by affinity chromatography/HPLC results in the rapid (two days) preparation of considerably more receptor protein. Such preparations have already proven essential in the generation of polyclonol antibodies directed towards the β_2-adrenergic receptor of frog erythrocytes (Strader *et al.*, 1984).

Purification of the β_1-adrenergic receptors of turkey red cell plasma membranes
As these techniques were found to be extremely successful for the purification of the β_2-adrenergic receptor, we sought to purify the β_1-adrenergic receptor from the turkey red blood cell. This was accomplished by essentially the same technique as utilized with the frog erythrocyte receptor protein. It is important to note, however, that β_1 receptors are present in turkey erythrocyte plasma membranes in lesser amounts when compared to the β_2 receptor in frog erythrocytes and that only 20–40% of these particulate β_1 binding sites can be solubilized with digitonin versus 80% for the β_2 receptor from frog erythrocyte preparations.

As shown in Table 4, crude digitonin extracts prepared from turkey red cell

Table 4. Purification of turkey erythrocyte β_1-adrenergic receptor by affinity chromatography and high performance liquid chromatography

Sample	Activity (pmol)[a]	Overall yield	Yield at each step	Specific activity (pmol[a]/mg protein)	Purification (fold) overall	step
Digitonin	119	100	100	1.15	1	1
Alprenolol Sepharose eluate	71.6	60	60	1108	963	963
First HPLC pass	62.1	52	87	4968	4320	4.5
Second HPLC pass	35.5	30	58	18 733	16 289	3.7

[a] As measured by [3H]DHA binding.

Typically, 100–200 pmol of soluble β-adrenergic receptor was prepared from 200–300 ml of purified turkey erythrocyte membranes with 300 ml of 2% digitonin, 100 μg/ml bacitracin, 10 μl/ml soybean tripsin inhibitor, 3×10^{-5} M phenylmethyl-sulfonyl fluoride, 3×10^{-4} M benzamidine, 100 mM NaCl/10 mM Tris-HCl pH 7.2 and applied to a column (200 ml) of alprenolol-Sepharose and eluted as described under Methods. Steric exclusion HPLC was as described in Fig. 3. Protease inhibitors were included until the first HPLC pass. Data shown here represents the average of two such preparations. Final receptor specific activities represent a 33 000-fold purification of receptor from membrane preparations.

membranes were chromatographed on Sepharose alprenolol affinity supports, and subsequently subjected to two sequential size exclusion HPLC runs. A typical size exclusion chromatogram run is shown in Fig. 3. This method results at an analytical level in the preparation of 2–5 μg of purified receptor protein from 100–200 ml of membranes. It is of interest that size exclusion chromatography of the turkey erythrocyte β_1 receptor displayed different elution properties from those of the frog β_2 receptor under the same conditions (Fig. 7) suggesting differences in protein structure between receptor subtypes.

When purified, β_1 preparations (i.e., one affinity chromatography and repetitive HPLC runs) were radiolabeled ($[^{125}I]$Na/chloramine T) and subjected to SDS–PAGE; autoradiography resulted in the visualization of two major bands: one at 45 000 M_r and one at 40 000 M_r, and in a ratio of 1:3–4 (Fig. 5). A trace band at 30 000 M_r was also visible, although preparation of receptor in the presence of protease inhibitors particularly EDTA (5 mM) markedly inhibited its occurrence. Using the photoaffinity probe $[^{125}I]$pABC, the same two polypeptides could be specifically labeled in membranes or in purified preparations of the β_1 receptor from the turkey red blood cell (Shorr *et al.*, 1982a).

Given the presence of two labelled proteins, the possibility exists that the

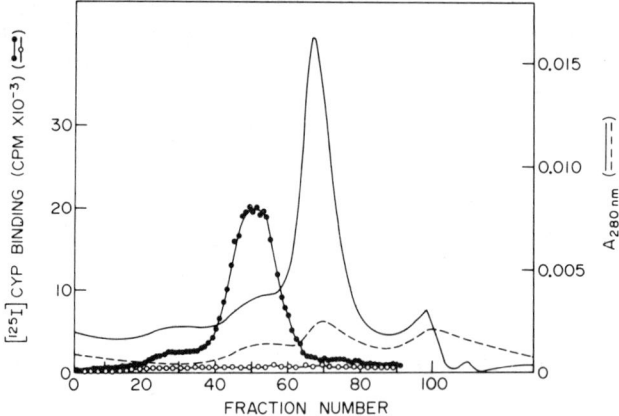

Figure 3. High performance steric exclusion chromatography of affinity purified turkey erythrocyte activity. Eluted fractions containing receptor activity after alprenolol-Sepharose chromatography (64.2 pmol) were pooled (80 ml) and concentrated to 2 ml by ultrafiltration using an Amicon concentration cell and PM-30 membrane. After concentration, the receptor (2 ml) was chromatographed on (2) TSK-4000 and (1) TSK-3000 steric exclusion columns tandem-linked (total volume = 72 ml). Solid line (——) denotes $A_{280\,nm}$. Receptor activity (●) (57 pmol) was located by chromatography of an aliquot of the affinity column eluate concentrate which had been incubated with $[^{125}I]$CYP (2 h 4°C) prior to chromatography. The open circles (○) represent receptor labeled with $[^{125}I]$CYP in the presence of 10^{-5} M alprenolol. Those fractions containing activity (fractions 43–60, 57 pmol) were pooled (5.4 ml) and concentrated by ultrafiltration to 2 ml prior to a second HPLC pass. Dashed line (– – –) denotes $A_{280\,nm}$ profile of the second HPLC pass. Total receptor activity recovered after two steric exclusion chromatography steps was 43 pmol. Mobile phase conditions were 0.2% digitonin 100 nm Tris-SO$_4$ pH 7.2, and 300 μl fractions were collected, 36 min after sample injection. From Shorr *et al.* (1982b).

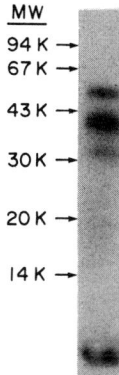

MW
94 K →
67 K →
43 K →
30 K →
20 K →
14 K →

Figure 4. SDS–PAGE of HPLC purified radiolabeled receptor activity. Those fractions containing receptor activity on steric exclusion chromatography of affinity column eluate concentrates were pooled and an aliquot (6 pmol) radiolabeled with Na[^{125}I] and chloramine T as described in Methods. After an additional HPLC run (to separate labeled receptor from labeled protein contaminants and from radiolabeled detergent), those fractions corresponding to receptor activity were pooled and an aliquot (25 000 cpm) subjected to SDS–PAGE and autoradiography (3 days exposure) as described in Methods. As shown, two major bands of radioactivity were revealed one at 40 000 M_r and one at 45 000 M_r; the ratio between the two bands was 3–4:1 for 40 000 M_r and 45 000 M_r peptides, respectively. Arrows to the right of the figure indicate the relative mobility of the various iodinated proteins used as M_r standards (cf. Methods). From Shorr *et al.* (1982b).

peptides visualized on SDS–PAGE are (1) separate receptor entities, (2) protomers of a single oligomeric complex, or (3) one protein is derived from the other by proteolysis. We have examined these possibilities by several criteria. Firstly, the two peptides were partially resolved by non-denaturing isoelectric focusing and by size exclusion HPLC (Shorr *et al.*, 1982a). Figure 6 shows an example of the separation of these two peptides by gel permeation HPLC. In this experiment a partially purified preparation was labeled with [^{125}I] pABC, chromatographed on HPLC and the fractions across the peak of receptor binding activity subjected to SDS–PAGE. The 45 000 M_r protein is found to be present primarily in the ascending fractions of the receptor binding activity peak (fractions 45–53); while the 40 000 M_r protein is found to be in the descending portion of the receptor peak (fractions 50–62). By pooling those fractions containing primarily one or the other protein and rechromatographing, it was possible to obtain isolated 40 000 M_r and 45 000 M_r peptide preparations. An example of such a preparation is shown in Fig. 7. Figure 7(A) shows the HPLC profile of the pooled fractions (42–48) and (56–59) from previous HPLC run as assayed by [^{3}H] DHA binding. In Fig. 7(B) is shown an autoradiogram of an SDS–PAGE of these same fractions from a preparation labeled with [^{125}I] pABC. These data demonstrate the homogeneity of each of the two receptor ligand binding proteins. The ability to separate these two peptides allowed their comparison as to the specificity of binding of adrenergic ligands and preliminary peptide mapping. Both 40 000 M_r and 45 000 M_r peptides were found to bind β-adrenergic

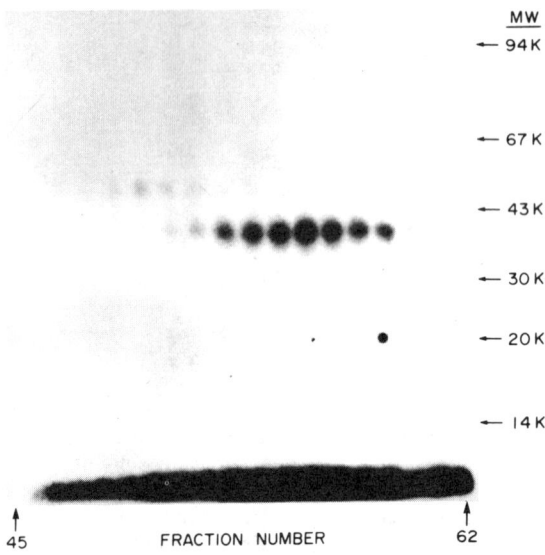

Figure 5. Resolution of the 40 000 M_r and 45 000 M_r polypeptides of purified receptor preparations by high performance liquid chromatography. A receptor preparation (51 pmol in 5 ml)
obtained by affinity chromatography and 1 HPLC pass was labeled with [^{125}I]pABC (16 nM
for 24 h at 4°C) and photolysed for 2.5 min using a mercury arc lamp 120 mm from the source.
After photolysis, the labeled receptor was concentrated by ultrafiltration to 2 ml and non-
incorporated [^{125}I]pABC was separated from labeled receptor by a second HPLC pass using
two 600 mm TSK-4000 and one 6000 mm TSK-3000 steric exclusion columns tandem-linked.
Mobile phase was 0.2% digitonin 100 mM Tris-SO$_4$ pH 7.2 and 300 μl fractions were
collected 36 min after sample injection. Aliquots (10 μl) of labeled fractions corresponding
to receptor activity (see Fig. 2) were then analyzed by SDS–PAGE. As shown, two labeled
proteins are visualized; one at 45 000 M_r being predominantly in fractions 45–52 (total 14 846
cpm) and one at 40 000 M_r being located predominantly in fractions 52–59 (total 51 467
cpm). From Shorr et al. (1982b).

ligands characteristically as assessed by competition of [^{125}I] cyanopindolol binding
with an identical typical β_1 specificity (Shorr et al., 1982a). Secondly, preliminary
partial proteolytic digestion of these two proteins in their folded globular states
labeled with either ^{125}I or [^{125}I] pABC revealed both similarities and differences in
the peptides generated as shown by SDS–PAGE (Shorr et al., 1982a). These data,
however, while predicting that the 40 000 M_r polypeptide was not simply derived
by proteolysis during the preparation from the 45 000 M_r protein prompted further
investigation into the relationship of structurally heterologous but pharmacologically non-distinct β_1 adrenergic receptors derived from a single tissue source.
These studies are primarily based on the preparation of detailed peptide finger-
prints from each of the two β_1 receptor forms. In order then to approach this
problem, we have developed techniques for more quantitative resolution of the
45 000 M_r and 40 000 M_r protein than could be obtained by gel permeation HPLC.
As a first step in this direction, we have examined the use of ion-exchange HPLC
matrices, primarily TSK–DEAE and Mono Q–DEAE columns (Pharmacia). Both

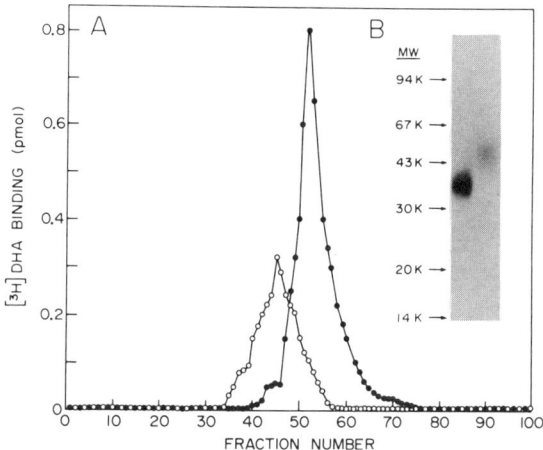

Figure 6. Preparation of purified 40 000 M_r and 45 000 M_r receptor polypeptides by high performance liquid chromatography. Purified receptor was prepared by affinity chromatography and an aliquot of the affinity eluate concentrate (9.3 pmol) radiolabeled with Na[^{125}I]/chloramine T. Both radiolabeled receptor and an aliquot of unlabeled activity were then chromatographed on steric exclusion columns as described in Fig. 2. Analysis of SDS–PAGE across each fraction of the ^{125}I-labeled receptor HPLC profile revealed a similar pattern to that shown in Fig. 5. Those fractions in the unlabeled receptor chromatography profile corresponding to either predominantly 40 000 M_r or 45 000 M_r receptor were then pooled (see Fig. 5, fraction 42–48 total 4 pmol; and fraction 56–59, total 7 pmol) and rechromatographed through the HPLC columns. (A) Receptor activity was located by [^{3}H] DHA (100 Ci/mmol) assay of each fraction and is represented as (●) for those fractions corresponding to the 45 000 M_r receptor pool (3.7 pmol) and as (○) for those fractions corresponding to the pooled 40 000 M_r receptor protein (5.4 pmol). In (B), 25.2 pmol of the same receptor preparation (affinity chromatography and 1 HPLC pass) were labeled with [^{125}I]pABC and subjected to a second HPLC pass and SDS–PAGE. Those fractions corresponding to the fractions of pooled unlabeled receptor in (A) were then pooled and subjected to SDS–PAGE on 10% acrylamide gels as described in Methods. Autoradiographs of dried gels were then prepared after 72 h exposure using Kodak XAR X-ray film and Kodak chemicals according to instructions for manual film processing. From Shorr *et al.* (1982b).

column types provide excellent resolution, although the pH range of the Mono Q gel exceeds that for the silica based TSK columns. The use of digitonin, however, in these systems served as a source of irreproducibility and short column life. More reproducible results were obtained by preparative SDS–PAGE and electroelution. Once separated from each other, each receptor protein was purified to homogeneity by fractionation on gel permeation TSK size exclusion columns equilibrated with a mixture of octylglucoside/SDS. In peptide mapping/SDS–PAGE experiments, a mixture of octylglucoside/SDS was found helpful for maximum enzyme digestion (Fred Marshall, Rob Shorr, unpublished observations). Figure 8(A) shows a calibration curve on these columns, using standard globular proteins ranging in molecular size

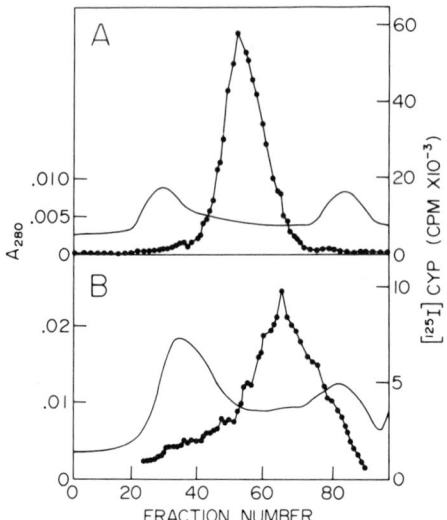

Figure 7. Separation of β_1- and β_2-adrenergic receptor subtypes by HPLC. (A) 2 ml of digitonin extract prepared from frog erythrocyte membranes as described in Shorr *et al.* (1981a, b) was chromatographed on two TSK-4000SW and one TSK-3000SW columns tandem linked as described in the legend to Fig. 1 except that 250 μl fractions were collected. (B) 2 ml of digitonin extract prepared from turkey erythrocyte membranes was chromatographed as in (A). For both (A) and (B) receptor activity (●) was located by assay with [^{125}I] cyanopindolol. Absorbance 280 nm is shown as (———).

from 94 000 M_r to 3500 M_r. Also shown are the results obtained when purified ^{125}I-labeled 45 000 M_r receptor protein was extensively digested with *Staphylococcus aureus* protease and chromatographed on the size exclusion (octylglucoside/SDS) HPLC columns. The starting material for the enzyme digest is shown as an inset in the figure. As shown, peptides were obtained ranging in apparent molecular size from 30 000 M_r to less than 1000 M_r. In order to generate a two dimensional fingerprint, each of the different peaks and shoulders of the digest chromatogram were subjected to isoelectric focusing (pH 3.5–10) on polyacrylamide gels polymerized with urea and Nonidet-40 a non-ionic detergent. As shown in Fig. 9, each of the different molecular weight regions in fact contains a number of peptides differing in molecular charge. The ability to prepare such detailed peptide maps using size exclusion HPLC/isoelectric focusing is essential in the biochemical comparison of receptor subtypes and structural sub-subforms.

DISCUSSION

Purification and characterization of hormone and drug receptors has been a difficult task primarily due to the extremely small quantities of these macromolecules in most cell types. β-Adrenergic receptors are closely coupled to the enzyme adenylate cyclase and have been the focus of considerable attention. Convenient model systems for study of these receptors are avian and amphibian erythrocyte plasma

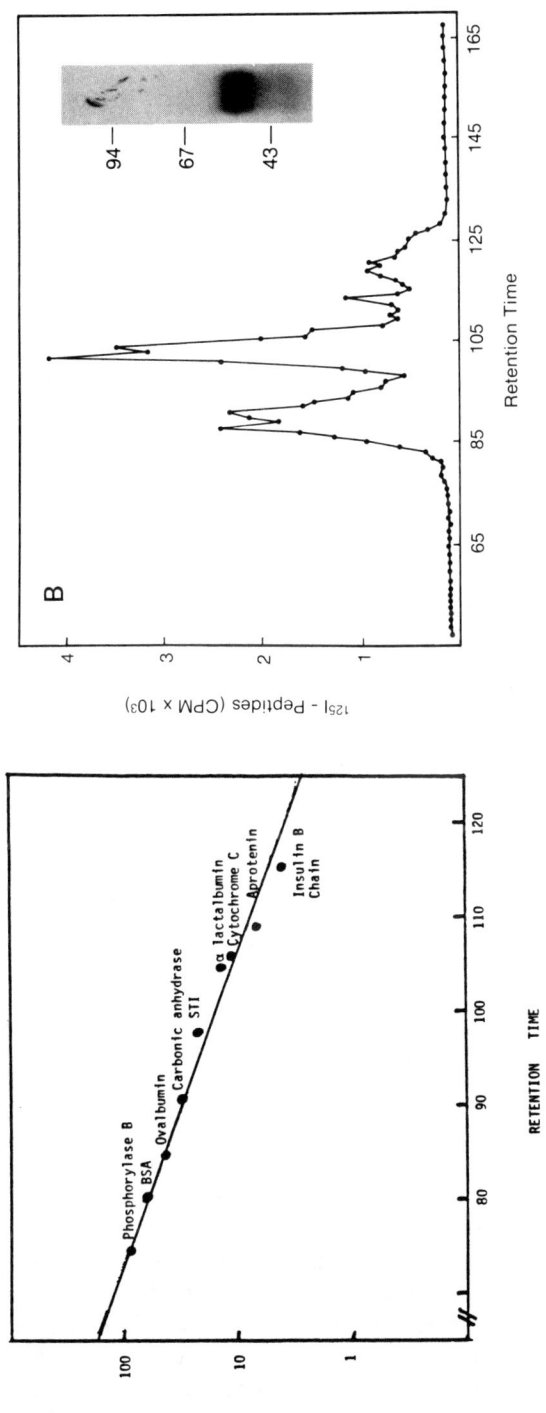

Figure 8. Molecular weight calibration of TSK–SW size exclusion columns equilibrated with octyl D-glucoside/SDS. (A) In order to estimate the molecular size of peptides generated from purified preparations of β receptor, a mixture of standard proteins (Pharmacia) to which had been added cytochrome c, aprotenin and insulin B chain (Pierce Chem Co.) was chromatographed. Columns were 6×300 mm and were one each tandem linked of TSK–SW 4000, TSK–SW 3000 and TSK–SW 2000. Mobile phase was 0.2% octyl D-glucoside, 0.08% SDS, 25 mM Tris-HCl, pH 6.8, 150 mM NaCl, 1 mM dithiothreitol. Flow rates were 0.2 ml/min. Absorbance was monitored at 220 and 280 nM using a Beckman Model 165 variable wavelength detector. Molecular weights are shown to the left of the figure versus retention time. (B) Purified β-adrenergic receptor containing 40 000 M_r and 45 000 M_r forms of the protein was prepared from turkey red blood cell membranes by affinity chromatography and non-denaturing size exclusion HPLC. [125I]-labeled purified 45 000 M_r protein was then obtained by ion exchange chromatography on a mono Q column eluted with a gradient to 0.5 M NaCl in 25 mM Tris-Cl pH 8.2 (data not shown). Purity was confirmed by SDS–PAGE of the radiolabeled fractions (inset). An aliquot of this material was then digested at 37°C for 40 min with a final concentration of 10 μg/ml *Staphylococcus aureus* protease A (Pierce). The peptide fragments were then separated on the size exclusion columns as described in (A). 0.2 ml fractions were collected and radiolabeled peptides located by gamma counting (Beckman 5500).

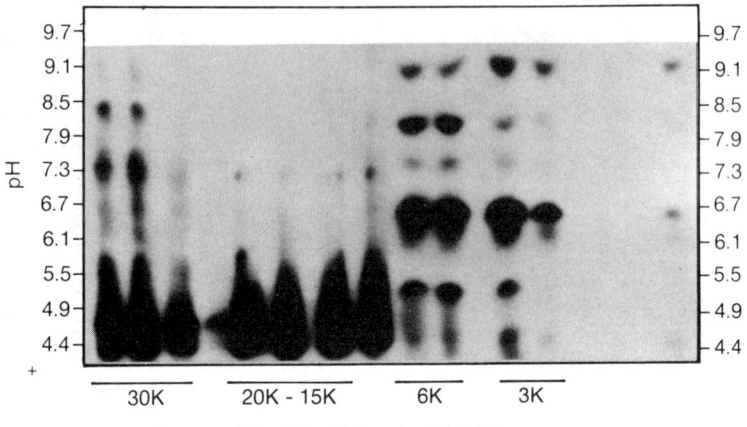

Apparent Peptide Molecular Weight

Figure 9. Second dimension isoelectric focusing of ^{125}I-labeled peptides generated by *Staph. areus* protease A digestion of ^{125}I-labeled 45 000 M_r β_1-adrenergic receptor. Fractions obtained by *Staphylococcus aureus* protease A digestion of purified ^{125}I-labeled 45 000 M_r β_1 receptor from turkey red blood cells as described in Fig. 8(B) were pooled according to apparent molecular size and aliquots subjected to isoelectric focusing as described in Methods. After focusing the gel was dried and subjected to autoradiography (5 days at $-70°$C) in order to visualize ^{125}I-labeled peptide fragments.

membranes which contain β_1- and β_2-adrenergic receptors, both which stimulate adenylate cyclase (Levitzki *et al.*, 1974; Mukherjee *et al.* 1976). These two subtypes of receptor are distinguished by differences in their pharmacological specificity for a variety of agonists and antagonists (Lands *et al.*, 1967). We have sought to address the question of a structural source of this pharmacological heterogeneity by purification of both β_1- and β_2-adrenergic receptors and their subsequent biochemical characterization. The data presented above clearly demonstrate the difference between β_1- and β_2-receptor subtypes from these sources by size exclusion HPLC and molecular identification of subunit components. However, more recent studies on the photoaffinity labeling of different β receptor subtypes in mammalian tissues suggest that the β_1- and β_2-adrenergic receptors in these mammalian systems may have more similarity in the size of their binding subunit. Both β_1 and β_2 receptors in these tissues appear to reside primarily on a 62 000–65 000 M_r peptide.

The use of size exclusion HPLC has been of fundamental importance in the rapid preparation of purified receptors and the demonstration (for the turkey erythrocyte) of structural heterogeneity within a pharmacologically homogeneous population of binding sites. It is apparent from the HPLC profiles shown that separation of receptor subtypes or subforms is probably not only due to size exclusion but also may reflect differential hydrophobic interactions with the silica matrix. The basis of this interaction is not immediately clear, particularly since the detergent digitonin is included in the mobile phase. Certainly, the use of size exclusion columns for final separation to biochemical homogeneity and a first dimension separation of peptide digests is extremely useful but brings forth a question as to why not direct application to reverse phase HPLC. We have examined reverse phase conditions for the β-adrenergic receptors using C-3, diphenyl, NH_2-silica, and C-18 column matrices. In all of these cases receptors were found to bind to the column supports but to be eluted only in trace amounts or without resolution. This problem seemed to be independent of the solvent system used or the column support. It is not unlikely that these problems reflect the protein hydrophobicity and more likely strong detergent interactions necessitating handling steps to remove detergents. These problems are currently being investigated in order to prepare purified peptides for protein sequencing. Nonetheless, techniques have been developed using size exclusion HPLC technology to resolve hydrophobic receptor membrane proteins as well as their digestion products in order to begin the detailed biochemical analysis of these important macromolecules.

REFERENCES

Benovic, J. L., Shorr, R. G. L., Heald, S. L., Lavin, T. N., Caron, M. G., and Lefkowitz, R. J. (1982). Purification and photoaffinity labeling of a mammalian β_2-adrenergic receptor. *Fed. Proc. 41*, 1161.

Carpenter, G. (1983). The biochemistry and physiology of the receptor kinase for epidermal growth factor (review). *Mol. Cell. Endocrinol. 31*, 1–19.

Caron, M. G., Srinivasan, Y., Pitha, J., Kiolek, K., and Lefkowitz, R. J. (1979). Affinity chromatography of the β-adrenergic receptor. *J. Biol. Chem. 254*, 2923–2927.

Caron, M. G. and Lefkowitz, R. J. (1976). Solubilization and characterization of the β-adrenergic receptor of frog erythrocytes. *J. Biol. Chem. 251*, 2374–2384.

De Lean, A., Hancock, A. A., and Lefkowitz, R. J. (1981). Validation and statistical analysis of a computer modelling method for quantitative analyses of radioligand binding data for mixtures of pharmacological receptor subtypes. *Mol. Pharmacol. 21*, 5–16.

Engel, G., Hoyer, D., Bertold, R., and Wagner, H. (1981). ($\pm$)[^{125}I]cyanopindolol, a new ligand for β-adreno receptors. Identification and quantitation of β-adreno receptor in guinea-pig. *Naunym Schmiedebergo Arch. Pharmacol. 317*, 277–285.

Gotti, C., Conti-tronconi, B. M., and Raftery, M. A. (1982). Mammalian muscle acetyleholine receptor purification and characterization. *Biochemistry 21*, 3148–3154.

Homcy, C. J., Rockson, S. G., Countaway, J., and Egan, D. A. (1983). Purification and characterization of the mammalian β_2-adrenergic receptor. *Biochemistry 22*, 660–668.

Lands, A. M., Arnold, A., McAuliff, J. P., Cuduena, F. P., and Brown, T. G. (1967). Differentiation of receptor systems by sympthomimetic amines. *Nature (Lond.) 214*, 597–598.

Lavin, T. N., Nambi, P., Heald, S. L., Jeffs, P. W., Lefkowitz, R. J., and Caron, M. G. (1982). [^{125}I]Para-azidobenzylcarazolol, a photoaffinity label for the β-adrenergic receptor characterization of the ligand and photoaffinity labeling of β_1- and β_2-adrenergic receptors. *J. Biol. Chem. 256*, 11944–11950.

Lefkowitz, R. J., Mukherjee, C., Coverstone, M., and Caron, M. G. (1974). Stereospecific [^{3}H](−)alprenolol binding sites, β-adrenergic receptors and adenyl cyclas. *Biochem. Biophys. Res. Commun. 69*, 703–710.

Levitzki, A., Atlas, D., and Steer, M. L. (1974). The binding characteristics and number of β-adrenergic receptors on the turkey erythrocyte. *Proc. Natl. Acad. Sci. USA 71*, 2773–2776.

Moore, H. P., Hartig, P. R., and Raftery, M. A. (1979). Correlation of polypeptide composition with functional events in acetylcholine receptor-enriched membranes from Torpedo, California. *Proc. Natl. Acad. Sci. USA 76*, 6265–6269.

O'Farrell, H. (1975(. High resolution two dimensional electrophoresis. *J. Biol. Chem. 250*, 4007–4021.

Raftery, M. A. Hunkapiller, M. W., Strader, C. D., and Hood, L. E. (1980). Acetylcholine receptor: complex of homologous subunits. *Science (Washington, DC) 208*, 1454–1457.

Rashidbaigi, A. and Ruoho, A. E. (1982). Photoaffinity labeling of β-adrenergic receptors: identification of the β receptor binding sites from turkey, pig and frog erythrocytes. *Biochem. Biophys. Res. Commun. 106*, 139–148.

Rodbell, M. (1980). The role of hormone receptors and GTP-regulatory proteins in membrane transduction. *Nature (Lond.) 284*, 17–22.

Shorr, R. G., Lyddiat, A., Lo, M. M. S., Dolly, J. O., and Barnadr, E. A. (1981a). Acetylcholine receptor from mammalian skeletal muscle: oligomeric forms and their subunit structures. *J. Biochem. 116*, 143–153.

Shorr, R. G. L., Lefkowitz, M. G., and Caron, M. G. (1981b). Purification of the β-adrenergic receptor: identification of the hormone binding subunit. *J. Biol. Chem. 256*, 5820–5826.

Shorr, R. G. L., Heald, S. L., Jeffs, P. W., Lavin, T. N., Strohsacker, M. W., Lefkowitz, R. J., and Caron, M. G. (1982a). The β-adrenergic receptor: rapid purification and covalent labeling by photoaffinity cross-linking. *Proc. Natl. Acad. Sci. USA 79*, 2778–2782.

Shorr, R. G. L., Strohsacker, M. W., Lavin, T. N., Lefkowitz, R. J., and Caron, M. G. (1982b). The β-adrenergic receptor of the turkey erythrocyte: molecular heterogeneity revealed by purification and photoaffinity labeling. *J. Biol. Chem. 257*, 12341–12350.

Shorr, R. G. L., Kempner, E. S., Strohsacker, M. W., Nambi, P., Lefkowitz, R. J., and Caron, M. G. (1984). Molecular size of β_1 and β_2-adrenergic receptors determined by radiation inactivation. *Biochemistry 23*, 747–753.

Stadel, J. M., Shorr, R. G. L., Limbird, L., and Lefkowitz, R. J. (1981). Evidence that a β-adrenergic receptor associated guanine nucleotide regulatory protein conveys GTP-8S dependent adenylate cyclase activity. *J. Biol. Chem. 256*, 8718–8723.

Stadel, J. M., DeLean, A., and Lefkowitz, R. J. (1982). Molecular mechanisms of coupling in hormone-receptor adenylate cyclase systems. *Adv. Enzymol. 53*, 1–43.

Strader, C. D., Pickel, V. M., Tong, T. H., Strohsacker, M. W., Shorr, R. G. L., Lefkowitz, R. J., and Caron, M. G. (1984). Antibodies to the β-adrenergic receptor: attenuation of catecholamine sensitive adenylate cyclase and demonstration of postsynaptic receptor localization in brain. *Proc. Natl. Acad. Sci. USA* (in press).

Progress in HPLC, Vol. 1, pp. 27–41
Parvez *et al.* (Eds)
© 1985 VNU Science Press

Determination of protein molecular weight by gel permeation chromatography equipped with low-angle laser light scattering photometer

TOSHIO TAKAGI

Institute for Protein Research, Osaka University, Yamadaoka, Suita, Osaka 565, Japan

INTRODUCTION

Determination of the molecular weight of proteins has been one of main subjects in the characterization of proteins. To discover protein molecular weights, researchers had relied heavily upon physicochemical techniques such as ultracentrifugal analysis, light scattering photometry and osmometry. The development of Sephadex 25 years ago brought about a big change in the situation. Andrews (1964) first showed that a unique relation exists between the retention times of most proteins and their molecular weights in gel permeation chromatography (GPC). Thereafter, biochemists became deeply committed to the use of 'molecular sieving effect' to determine protein molecular weights. Polyacrylamide gel electrophoresis in the presence of sodium dodecyl sulfate (SDS) is the most widely used among the techniques. Recent advances in GPC in HPLC mode made the procedure of molecular weight determination more reproducible and efficient. Despite the widespread use of the techniques of protein molecular weight determination that depend on the molecular sieving effect, the possible range of the techniques is limited. One of the most important premises of these techniques is that a unique relation exists between protein molecular weights and their sizes determining the retention times. Such a relation, however, is expected to hold only for a group of proteins which are homologous among them with respect to chemical structure and molecular conformation. Often it is impossible to foresee whether a sample protein is homologous to the proteins used as standards in the above context.

Classical physicochemical techniques require a large sample volume which is often difficult to afford. There has been, therefore, the need for a method which is not only as convenient as the molecular sieving techniques but also as accurate as the classical physicochemical techniques. Only a single technique satisfying this is now emerging. In this technique a measuring system is used

---

Abbreviations: GPC, gel permeation chromatography; LALLS, low-angle laser light scattering; GPC/LALLS, monitoring of elution from a GPC column using a LALLS photometer and supplementary equipment.

which is composed of a GPC system, a low-angle laser light scattering (LALLS) photometer, and supplementary equipment. This chapter is an introduction to the relatively recent protein characterization technique called the GPC/LALLS technique for convenience.

HISTORICAL BACKGROUND

Conventional light scattering technique

Debye (1944) first suggested that the molecular weight of a polymer can be determined by the measurement of scattered light. However, Putzeys and Brosteaux (1935) had indicated that the light scattering intensities of several proteins were closely correlated to their molecular weights. Halwer *et al.* (1951) demonstrated that the molecular weights of several water-soluble proteins could be evaluated precisely by the light scattering method. The experimental procedure in such light scattering measurements required highly experienced technicians and a well-equipped laboratory. The technique, therefore, gave way to the ultracentrifugal technique, and the latter then to the molecular sieving techniques as described earlier. A chapter in the textbook by Tanford (1961) contains an excellent description of the outline of the conventional light scattering technique, and it is also helpful for understanding the basis of the GPC/LALLS technique.

Light scattering detector

With the development of gel permeation chromatography, various kinds of conventional measuring instruments, such as the spectrophotometer and differential refractometer, were modified so that they could be used to monitor elution of polymers from a GPC column. Even the light scattering photometer was used: Kaye *et al.* (1971) developed a scattering photometer with unique optics – the low-angle laser light scattering (LALLS) photometer. As will be described below, this scattering photometer had a feature suitable for the use as a detector. Kato *et al.* (1979) also developed a scattering photometer of the detector type which monitored scattered light at 90°. However, this type of scattering photometer failed to find its way as a commercial product. Immediately after the development of the LALLS photometer, Ouano and Kaye (1974) demonstrated the high performance of their instrument in the detection of polystyrene eluted from a GPC column using chloroform as the solvent.

Recent development of high performance GPC columns adequate for the use of aqueous solvents opened the use of the scattering photometer for monitoring of elution of proteins from such columns. Fukutomi *et al.* (1980) illustrated that molecular weights of proteins can be determined efficiently by monitoring their elution from a TSK-GEL SW GPC column. Takagi (1981, 1982) further assessed the performance of the technique. These studies stimulated application of the technique in the field of biochemistry.

Commerical models of low-angle laser light scattering photometer

The LALLS photometer developed by Kaye *et al.* (1971) developed into the KMX-6 Low-angle Laser Light Scattering Photometer (Chromatix Co., Sunnyvale,

Figure 1. Photograph of the Low-angle Scattering Photometer of TSK Model LS-8000.

Calif.). This photometer is an excellent instrument designed to make possible the measurement of scattered light from a liquid (Kaye and Havlik, 1973). The KMX-6 scattering photometer is too sophisticated and expensive to be used as a detector in GPC. The TSK LS-8 Low-angle Scattering Photometer (Toyo Soda Co., Tokyo, Japan) (Fukuda *et al.*, 1979) and the LSD-100 Low-angle Laser Light Scattering Photometer (Chromatix Co.) were developed with the aim of building a GPC detector. Figure 1 shows the external appearance of the most recent model of TSK Low-angle Laser Light Scattering Photometer (LS-8000).

THEORETICAL BACKGROUND

Debye (1944) interpreted the light scattering from a polymer solution as the result of the fluctuation of polymer concentration. According to his theory, the fluctuation makes the polymer solution inhomogeneous with respect to the refractive index from a microscopic point of view, to make an impinging light deviate from a straight path. This way of thinking applies to particles small enough relative to the wavelength of the impinging light. The size of the particles analyzed in the GPC satisfies this premise. We will, therefore, consider the light scattering phenomenon according to Debye's theory.

Let us assume that we are observing a local region with an extremely small volume, v, in a polymer solution with weight concentration of c_0. When the concentration fluctuates to c, the amplitude of the light scattered from the region is proportional to v and the change in the refractive index produced there. The latter is equal to the product of the degree of the fluctuation in the polymer concentration, $c - c_0$, and the specific refractive index increment of the polymer, $(\mathrm{d}n/\mathrm{d}c)$. The intensity of the scattered light is proportional to the square of the amplitude, and, therefore

$$\text{(intensity of the scattered light)} \propto v^2 \langle (c - c_0)^2 \rangle (\mathrm{d}n/\mathrm{d}c)^2. \tag{1}$$

We actually observe the time-averaged intensity of the scattered light. This context is expressed in Equation (1) by enclosing the square of the concentration change by

angular brackets. According to the theory of fluctuation, the angle-bracketed value
is expressed as follows:

$$\langle (c - c_0)^2 \rangle \propto cv^{-1}(\partial\pi/\partial c)^{-1}. \tag{2}$$

The appearance of $(\partial\pi/\partial c)$, the change in osmotic pressure with concentration, may
appear peculiar, but it will be accepted as quite normal if considered as follows:
(1) we have a polymer solution, and a part of which is enclosed in a cylinder being
separated from the bulk of the solution by a semi-permeable piston; (2) if the piston
is moved inward to produce an increase of the polymer concentration, osmotic
pressure will be applied to the piston in proportion to the change in concentration;
(4) the change in osmotic pressure with concentration is, therefore, a determining
factor in the work necessary to produce a fluctuation; and (5) thus the reciprocal
of the change of osmotic pressure with concentration can be taken as a measure of
the probability of the spontaneous fluctuation.

Osmotic pressure can be expressed as follows:

$$\pi = RT(c/M + Bc^2 + \cdots). \tag{3}$$

The derivative, $(\partial\pi/\partial c)$, in Equation (2) can, therefore, be expressed as

$$(\partial\pi/\partial c) = RT(1/M + 2Bc + \cdots). \tag{4}$$

If the light scattering is measured for a polymer solution at an enough low concentration, its intensity can be expressed as

$$\text{(intensity of scattered light)} \propto vMc(\mathrm{d}n/\mathrm{d}c)^2. \tag{5}$$

If a light scattering photometer, with high sensitivity and with a flow-through
type cell having a small internal volume is available, it will function as a detector
which gives an output

$$\text{(output)}_{LS} \propto Mc(\mathrm{d}n/\mathrm{d}c)^2. \tag{6}$$

A low-angle laser light scattering photometer is an instrument giving such an output.
As described above, the light scattering study is correlated closely with osmometry.
The sample solution in the former is required to be in equilibrium with the solvent
used, in the same manner as in the latter. When a sample is applied to a LALLS
photometer via a GPC column, the equilibrium is established during the passage
through the column.

OUTLINE OF A LOW-ANGLE LASER LIGHT SCATTERING PHOTOMETER

Figure 2 shows a schematic picture of the optics of the LALLS photometer. The
first part of the optics makes well-focused laser light (He–Ne) impinge upon the
cell. The second part, shown in Fig. 3 in three dimensions to facilitate understanding, has the following features: (1) a pair of long windows makes the air/glass
boundaries apart from the cell so that the scattered light is not contaminated from
the stray light coming from the boundaries which are major sources of such
light; (2) a cell with a very small volume (ca. 10 μl) makes it possible to use a
LALLS photometer as a detector to monitor elution of proteins from a GPC

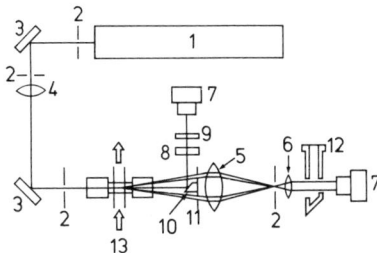

Figure 2. Outline of the optics of the Low-angle Scattering Photometer of TSK Model LS-8000. 1 He−Ne laser, 2 apertures, 3 mirrors, 4 condensing lens, 5 relay lens, 6 detector lens, 7 photomultipliers, 8 filter, 9 diffuser, 10 photo-trap, 11 annulus, 12 microscope, 13 flow-cell.

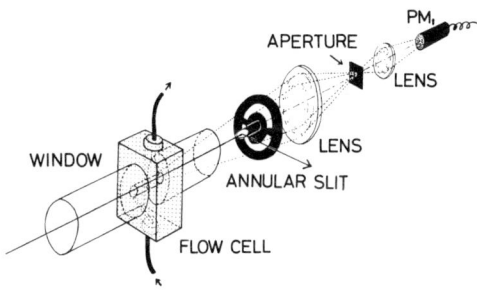

Figure 3. Schematic diagram illustrating how the scattering light is detected at a low-angle in the instrument shown in Fig. 1.

column; (3) the optics is designed so that only a minute volume [v in Equation (5)] as small as the order of 0.1 μl in the cell is observed, and consequently a dust particle or an air-bubble intruding into the cell gives a spike-like noise which is easily recognized; (4) the annular slit effectively cuts the impinging light and collects the scattered light in a high yield and at a low-angle around 5°.

The ingenious design of the optics further brings about the following advantages: (1) the measurement at a low-angle makes it less likely that scattered light from different points interfere with each other inter- or intra-molecularly, and this makes extrapolation neither to zero protein concentration nor to zero angle necessary; (2) the use of a LALLS photometer in a mode of detector with a flow-through cell makes it possible to eliminate dust particles in the scattering photometry by providing a membrane-filter (UMF) in front of the scattering cell. As shown in Fig. 2, the scattered light collected through the annular slit is monitored by the photomultiplier (PM_1), and its intensity is compared with that of the impinging light monitored by another photomultiplier (PM_2).

The other two models of low-angle laser light scattering photometer have optics with designs somewhat different from those of the model mentioned above. All three models are, however, descendants of the prototype reported by Kaye *et al.* (1971), and are similar in their major features.

OUTLINE OF INSTRUMENTATION

A low-angle laser light scattering photometer functions fully only when incorporated into a high-performance gel chromatography system. Figure 4 shows the measuring system being used in the author's laboratory, as an example.

Each of the components in Fig. 4 will be interpreted in order along the flow path. The reservoir (R) has a volume of 3.5 l and is filled with a suitable buffer solution which has been filtered through an ultrafilter. Preferably it contains 0.02% sodium azide to prevent bacterial growth. The buffer solution is suctioned via the filter (F_1) made of sintered stainless steel. Air dissolved in the buffer solution is eliminated during the passage through the degasser (DG, Elma Optical Works, Model ERC-3310) having a long tubing with a wall permeable only to gaseous materials and enclosed in a chamber maintained at moderately negative pressure. Supply of the buffer solution at a constant speed is controlled by a solvent delivery system with a sample injector (HLC, Toyo Soda Co., High Speed Liquid Chromatograph, Model 803D). Between the dual plunger pump and the sample injector, a coiled stainless steel tubing (Co, 0.1 mm × 2 m) was installed to build-up back-pressure required for the pump to function properly. In front of the sample injector, a pressure gauge (G) and another sintered stainless steel filter (F_2) is installed. The flow then goes through a guard column (GC; Toyo Soda Co., TSK-GEL GSWP, 7.5 × 100 mm) and a main column (C; Toyo Soda Co., TSK-GEL G3000SW, 7.5 × 600 mm). Elution from the column is monitored by the following three kinds of detectors connected in series: a u.v. spectrophotometer (UV; Toyo Soda Co., TSK Model UV-8 II), a precision differential refractometer (DR; Toyo Soda Co., TSK Model RI-8) and a low-angle laser light scattering photometer (LS; Toyo Soda Co., TSK Model LS-8). The outputs of the detectors are registered by a recorder (Re).

ELUTION CURVES OF BOVINE SERUM ALBUMIN

Figure 5 shows three elution curves obtained for bovine serum albumin by the

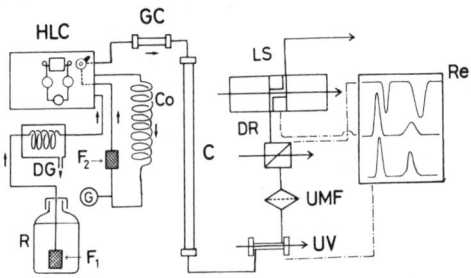

Figure 4. Outline of instrumentation used in the author's laboratory to determine protein molecular weights by the GPC/LALLS technique (for details, see text). The elution curves shown in this article were actually obtained using the measuring system in which the equipment behind the column were connected in the following order: UV–UMF–LS–RI. The order was subsequently changed to that shown in Fig. 4 in order to minimize the effect of disturbance of flow in the light scattering cell having internal volume significantly larger than the others.

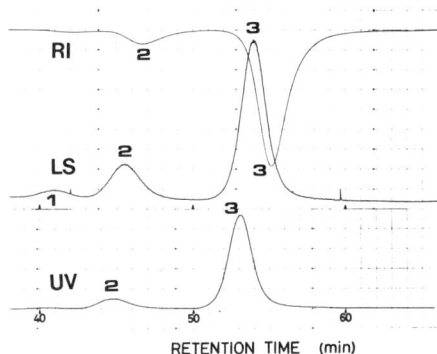

RETENTION TIME (min)

Figure 5. Elution curves of bovine serum albumin. The three curves are, from bottom to top, the tracings obtained by the spectrophotometer (UV), the LALLS photometer (LS) and the refractometer (RI), respectively. Bovine serum albumin (Armour, 5 × crystallized), 0.4 mg in 100 μl; solvent used for both column equilibration and sample preparation, 0.025 M NaH_2HPO_4–0.075 M Na_2HPO_4, pH 7 containing 3 mM NaN_3; flow rate, 0.31 ml/min; temperature, 25°C (r.t.); gain settings of the detectors, UV 0.64, LS 32, and RI 64.

measuring system shown in Fig. 4. They are helpful to illustrate features of the output of the LALLS photometer in comparison with those of the others. Displacement observed among positions of the corresponding peaks is due to the difference in positions of the detectors along the flow path and the pens on the recorder charts. For proteins without molecular weight distribution, the height of each peak can be taken as a measure of the output of the corresponding detector.

Peaks designated as 2 and 3 can be assigned to dimer and monomer of bovine serum albumin, respectively. The two naturally share the same value of (dn/dc) and extinction coefficient. For either of the tracings by the refractometer and the spectrophotometer, the ratio of the height of peak 2 to peak 3 is 1:10. On the other hand, the ratio in the tracing of the scattering photometer is 1:5. The above difference in the output ratios clearly exemplifies that the LALLS photometer functions as a detector giving output proportional to the product of molecular weight and weight concentration. Peak 1 can be assigned to a mixture of materials eluted at the void volume of the column. Aggregates of bovine serum albumin may be eluted at the position, but the deflection in the refractometer output, more sensitive than that in the spectrophotometer, suggests that materials (often called 'micro gel') other than the aggregates are included.

MOLECULAR WEIGHT DETERMINATION OF SIMPLE PROTEINS

Figure 6 shows the elution curves obtained for a mixture of four simple proteins of which the molecular weights range between 20 000 and 300 000. Such proteins, consisting of amino acids only, share almost the same value of (dn/dc), unless one is heavily biased with respect to amino acid composition. From Equation (5) it is expected that the intensity of scattered light is proportional to cM for

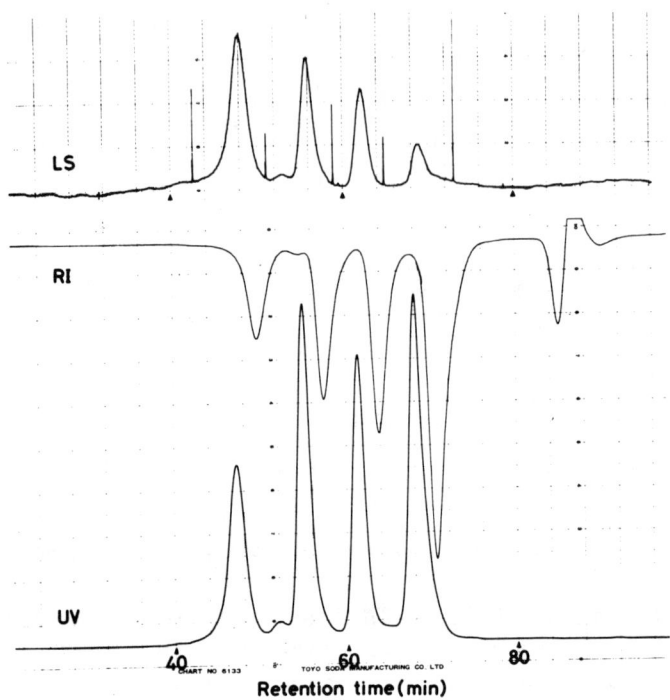

Figure 6. Elution curves of a mixture of standard proteins. The curves are identified by the same symbols as in Fig. 5. Major peaks are assingnable from left to right, to glutamic dehydrogenase (280 000, 60 μg), lactic dehydrogenase (142 000, 90 μg), enolase (88 000, 100 μg) and adenylate kinase (21 500, 130 μg). Sources of the proteins are yeast, except porcine heart for the dehydrogenase. A pair of numbers in each parenthesis are molecular weight and amount contained in the sample solution of 100 μl, of respective protein. Solvent, the same as that in Fig. 5, except the addition of 0.2 M NaCl; flow rate, 0.31 ml/min; temperature, 25°C (r.t). Gain setting; UV 1.28, RI 32, and LS 32.

such proteins. Output of the refractometer, $(output)_{RI}$ shown in Fig. 4 can be expressed as

$$(output)_{RI} \propto (dn/dc)c. \tag{7}$$

For simple proteins, $(output)_{RI}$ is, therefore, expected to be proportional to c. Thus, it can be expected that a linear relation exists between the output ratios, $(output)_{LS}/(output)_{RI}$ and protein molecular weights. In Fig. 7, the output ratios for the proteins shown in Fig. 6 are plotted versus their molecular weights. The ratios were also measured with a mixture of four of proteins, which are or closely resemble simple proteins, are also included in Fig. 7. The plots show a straight line which crosses the origin. This plot can be used as a calibration line from which molecular weight of such a protein can be uniquely determined from the output ratios measured for the proteins. The elution curve obtained by the u.v. spectrophotometer is of no use for molecular weight determination in this case.

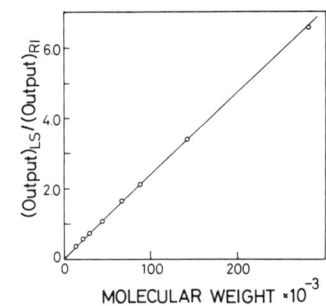

Figure 7. Calibration line for molecular weight determination of simple proteins.

According to the preliminary results I obtained the mode of molecular weight determination described above can also be carried out in either in 8 M urea or 6 M guanidine hydrochloride. Thus, the technique offers reliable molecular weights for subunits of proteins as well as for their parent oligomeric proteins. In the case of 6 M guanidine hydrochloride, we used a Milton-Roy single plunger pump instead of the highly sophisticated pumping system shown in Fig. 4 to avoid any possible damage by corrosion.

MOLECULAR WEIGHT DETERMINATION OF A GLYCOPROTEIN

In this section the method of estimating the molecular weight of α_1-acid glyco-protein using the measuring system shown in Fig. 4 (Maezawa and Takagi, 1983) will be described. It is intended to show how the molecular weight of a protein with a non-amino acid component can be estimated by the present technique. The polypeptide moiety of the glycoprotein has been shown to have molecular weight of 21 270 by sequence study (Schmid *et al.*, 1973). The carbohydrate content of the protein was estimated to be 44.1% (w/w) (Li *et al.*, 1983). Thus, the molecular weight of the protein can be estimated to be 38 000. Common value of (dn/dc) could be properly assumed for simple proteins. In the case of glycoproteins rich in carbohydrates, the assumption is no longer valid. An alternative procedure was, therefore, taken to evaluate molecular weight of α_1-acid glycoprotein. The output of the u.v. spectrophotometer, included in the measuring system shown in Fig. 4, can be expressed as

$$(\text{output})_{UV} \propto Ec \qquad (6)$$

where E is extinction coefficient on the weight basis. From Equations (5)–(7) the molecular weight of a protein can be expressed as

$$M \propto (\text{output})_{LS}(\text{output})_{UV}(\text{output})_{RI}^{-2}E^{-1}. \qquad (7)$$

Figure 8 shows the elution curves obtained for the glycoprotein. The follow-ing proteins were used as standards: bovine serum albumin (66 267, 0.670 ml mg^{-1} cm^{-1}), hen's ovalbumin (42 700, 0.735), bovine carbonic anhydrase (29 000, 1.90), and bovine pancreatic ribonuclease (13 700, 0.706) where the two

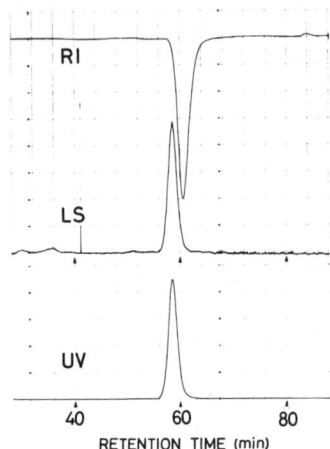

Figure 8. Elution curves of human α_1-acid glycoprotein. Symbols, the same as in Fig. 5. Solvent, the same as in Fig. 6; sample, 0.5 mg of the glycoprotein in 100 μl of the solvent; flow rate, 0.33 ml/min; gain setting, UV 1.28, LS 32, and RI 32.

values in each parenthesis are molecular weight and extinction coefficient at 280 nm on the weight basis, respectively. Figure 9 shows the elution curves obtained for a mixture of the standard proteins. Figure 10 shows the plots of the right side of Equation (7) for the data shown in Fig. 9 versus their molecular weights. A straight line was obtained to give a calibration line. Using the reported value of 0.881 ml mg^{-1} ml^{-1} at 280 nm for the extinction coefficient of the glycoprotein, its molecular weight was estimated to be 38 000 ± 400 ($n = 4$). This is in excellent agreement with the value of 38 000 determined from the sequence data and the carbohydrate content.

EVALUATION OF THE GPC/LALLS TECHNIQUE

Various techniques have been applied to the determination of the molecular weight of human α_1-acid glycoprotein, and the values obtained are compiled in Table 1. The results show that the techniques that depend on the molecular sieving effect cannot be applied to the glycoprotein with high content of non-polypeptide component. The results described above clearly demonstrate that the GPC/LALLS technique described in this chapter can be used to determine the molecular weights of proteins to which no techniques depending on the molecular sieving effect can be applied.

The GPC/LALLS technique requires some experience in the operation of the system that is shown in Fig. 4. Once one has become acquainted with the system, a series of measurements can be finished within several hours. For simple proteins, the technique is extremely convenient. For a conjugated protein whose extinction coefficient is unknown, the determination of the coefficient will require more time than that required for the main experiments. To make the procedure both time- and sample-saving, the quantitative amino acid analysis might be the best choice.

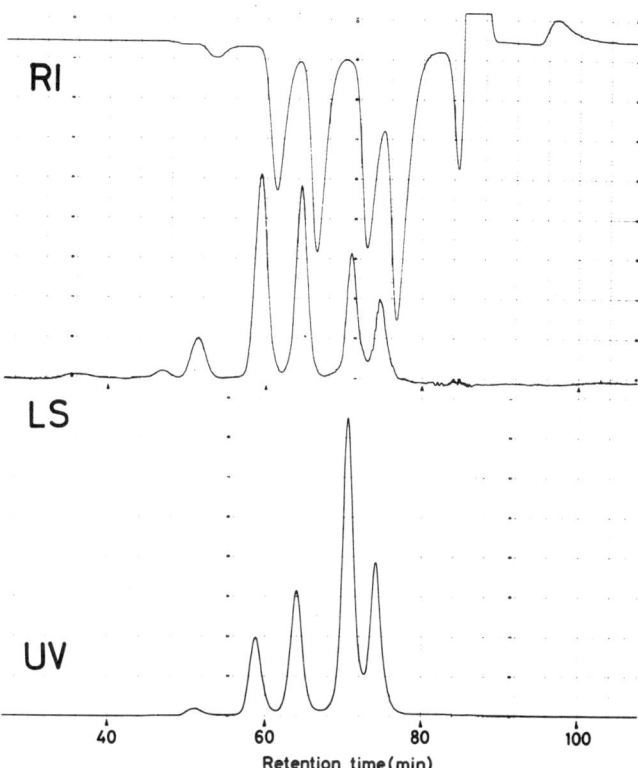

Figure 9. Elution curves of a mixture of standard proteins. Symbols, the same as in Fig. 5. Major peaks are assignable to, from left to right, dimer of bovine serum albumin (major for the scattering tracing only), bovine serum albumin (66 267; 0.670 ml mg^{-1} cm^{-1}; 330 μg), hen's ovalbumin (42 700; 0.735; 400 μg), bovine carbonic anhdrase (29 000; 1.90; 500 μg) and bovine pancreatic ribonuclease (13 700; 0.706; 610 μg). Numbers in each parenthesis are molecular weight, extinction coefficient with dimension described in the first parentheses, and amount contained in the sample solution of 100 μl for the respective protein. Solvent, the same as in Fig. 6; flow rate, 0.31 ml/min; gain setting, the same as in Fig. 8.

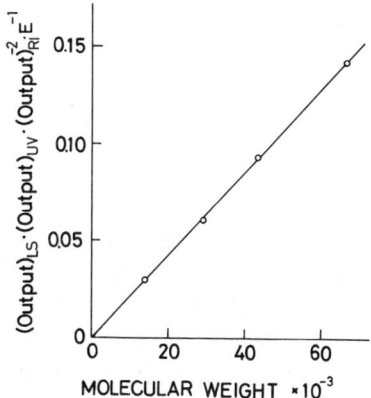

Figure 10. Calibration line for molecular weight determination of human α_1-acid glycoprotein.

T. Takagi

Table 1. Molecular weights of human α_1-acid glycoprotein estimated by various techniques

Technique	Molecular weight estimated	Reference
Amino acid sequence and carbohydrate content	38 000	Schmid et al. (1973) and Li et al. (1983)
GPC	70 000	Kawasaki et al. (1966)
GPC in the presence of denaturants	31 500 32 000	Ui (1981) Leach et al. (1980a, b)
SDS-polyacrylamide gel electrophoresis	40 000–53 000 (depending on gel concentration)	Leach et al. (1980a)
Classical physicochemical method	39 000–41 000	Schmid et al. (1975)
GPC/LALLS	38 000	Maezawa and Takagi (1983)

The GPC/LALLS technique can be applied to any protein which can be applied to presently available high-performance gel chromatography columns. Exceptions are proteins which absorb in the visible region. Absorption of light causes malfunction in both of the detectors, especially in the refractometer using a white light. The situation can be improved by changing the light source of the refractometer to one with wavelength characteristics similar to a He–Ne laser. Proteins which absorb light at or near the wavelength of a He–Ne laser will still be outside the range of the GPC/LALLS technique. The sedimentation equilibrium technique is of value as an alternative. The above two techniques will continue to coexist as the major ones complimentary with each other in the field of physicochemical characterization of proteins.

Until now, the application of the GPC/LALLS technique to studies of proteins has attracted interest of a limited number of groups. To my knowledge the number of papers dealing with this technique is ten or so, and they are compiled in Table 2. It is to be noted that the technique has been successfully applied to membrane proteins solubilized by surfactants as is clear from Table 2.

CONCLUSIONS

The HPLC techniques are now being welcomed enthusiastically in the field of biochemistry. In this book, various kinds of the techniques are introduced. Among them, the GPC/LALLS technique described in this article seems to be least known. As is clear from Table 2, the technique has been applied to various objects which surely attract biochemists. It is hoped that more and more people in the field will become interested in the GPC/LALLS technique.

Table 2. Chronology of the application of the GPC/LALLS technique to proteins

Protein	Purpose	Special additive to aqueous buffer	Reference
Typical water-soluble proteins	Assessment		Fukutomi *et al.* (1980)
Asp. oryzae α-amylase	Molecular weight		Takagi (1981)
Thyroglobulin	Molecular weights	6 M guanidine hydrochloride	Takagi *et al.* (1981)
E. coli porin	Molecular weights (oligomer and subunit)	2 mM SDS	Kameyama *et al.* (1982) and Ishii *et al.* (1983)
Torpedo acetylcholine receptor	Molecular weight (oligomer)	0.1% sodium deoxycholate	Imamura *et al.* (1982)
Bovine lens proteins	Component analysis		Bindels *et al.* (1982)
Typical water-soluble globular proteins	Assessment		Takagi (1983)
Human α_1-acid glycoprotein and hen's ovomucoid	Assessment		Maezawa and Takagi (1983)
E. coli porin and λ-receptor protein	Molecular weights (oligomers)	0.1% octaethylene glycol *n*-dodecyl ether ($C_{12}E_8$)	Maezawa *et al.* (1983) and Ishii *et al.* (1983)
Cannine kidney Na^+,K^+-ATPase	Molecular weights (oligomers)	0.1% $C_{12}E_8$	Hayashi *et al.*
Equine kidney Na^+,K^+-ATPase	Molecular weights (oligomer)	0.01% $C_{12}E_8$	Nakao *et al.* (1983a
Torpedo acetylcholine receptor	Molecular weights	0.2% SDS	Imamura *et al.* (1983)
Hen's ovalbumin	Aggregate formation on heat denaturation		Kato *et al.* (1983)

ACKNOWLEDGEMENTS

This work was supported by Scientific Research Funds (grant nos. 321930, 447123 and 56580110) of the Ministry of Education, Science and Culture of Japan. The author expresses his thanks to Dr Shigenori Maezawa for his cooperation in the research. He also thanks the people of the HPLC Developing Groups of Toyo Soda Company for their cooperation in the improvement of the measuring system.

REFERENCES

Andrews, P. (1964). The gel-filtration behaviour of proteins related to the molecular weights over a wide range. *Biochem. J. 96*, 595–606.

40 *T. Takagi*

Bindels, J. B., de Man, B. M., and Hoenders, H. J. (1982). High-performance gel chromatography of bovine eye lens proteins in combination with low-angle laser light scattering. *J. Chromatogr. 252*, 255–267.

Debye, P. (1944). Light scattering in solution. *J. Appl. Phys. 15*, 338–342.

Fukuda, M., Fukutomi, M., Banba, N., and Hashimoto, T. (1979). Construction of a low-angle laser light scattering photometer and examination of its performance. *Toyo Soda Res. Rep. 23*, 110–118 (in Japanese).

Fukutomi, M., Fukuda, M., and Hashimoto, T. (1980). Evaluation of water-soluble polymers by GPC-light scattering measurement. *Toyo Soda Res. Rep. 24*, 33–41 (in Japanese).

Halwer, M., Nutting, G. C., and Brice, B. A. (1951). Molecular weight of lactoglobulin, ovalbumin, lysozyme and serum albumin by light scattering. *J. Am. Chem. Soc. 73*, 2786–2790.

Hayashi, Y., Takagi, T., Maezawa, S., and Matsui, H. (1983). Molecular weights of $\alpha\beta$-protomeric and oligomeric units of soluble (Na^+,K^+)-ATPase determined by low-angle laser light scattering after high-performance gel chromatography. *Biochim. Biophys. Acta. 748*, 153–167.

Imamura, T., Konishi, K., and Konishi, K. (1982). Physico-chemical properties of acetylcholine receptor proteins from *Narke japonica. J. Biochem. 92*, 1901–1910.

Imamura, T., Konishi, K., and Konishi, K. (1983). Molecular weight of acetylcholine receptor subunits from *Narke japonica. J. Biochem. 94*, 917–923.

Ishii, J. N., Takagi, T., Kameyama, K., and Nakae, T. (1982). The use of the laser light scattering for the molecular weight determination of membrane proteins. *Tokai J. Exp. Clin. Med. 7 (Suppl.)*, 157–164.

Kato, T., Kanda, A., Takahashi, A., Noda, I., Maki, S., and Nagasawa, M. (1979). Determination of the degree of branching of comb-shaped branched polystyrenes by a GPC/LS method. *Polymer J. 11*, 578–581.

Kameyama, K., Nakae, T., and Takagi, T. (1982). Estimation of molecular weights of membrane proteins in the presence of SDS by low-angle laser light scattering combined with high-performance porous silica gel chromatography. Confirmation of the trimer structure of porin of the *E. coli* outer membrane. *Biochim. Biophys. Acta. 706*, 19–26.

Kato, A., Nagase, Y., Matsudomi, N., and Kobayashi, K. (1983). Determination of molecular weight of soluble ovalbumin aggregates during heat denaturation using low-angle laser light scattering technique. *Agr. Biol. Chem. 47*, 1829–1834.

Kawasaki, T., Koyama, J., Yamashina, I. (1966). Isolation and characterization of α_1-acid glycoprotein from rat serum. *J. Biochem. 60*, 554–560.

Kaye, W. and Havlik, A. J. (1973). Low-angle laser light scattering–absolute calibration. *Appl. Opt. 12*, 541–550.

Kaye, W., Havlik, A. J., and McDaniel, J. B. (1971). Light scattering measurments on liquids at small angles. *Polym. Lett. 9*, 695–699.

Leach, B. S., Collawn, Jr, J. F., and Fish, W. W. (1980a). Behavior of glycopolypeptides with empirical molecular weight estimation methods. 1. In sodium dodecyl sulfate. *Biochemistry 19*, 5734–5741.

Leach, B. S., Collawn, Jr, J. F., and Fish, W. W. (1980b). Behavior of glycopolypeptides with empirical molecular weight estimation methods. 2. In random coil producing solvents. *Biochemistry 19*, 5741–5747.

Li, Z.-Q., Perkins, S. J., and Loucheux-Lefebvre, M. H. (1983). α_1-Acid glycoprotein: a small-angle neutron scattering study of a human plasma glycoprotein. *Europ. J. Biochem. 130*, 275–279.

Maezawa, S., Hayashi, Y., Nakae, T., Ishii, J., Kameyama, K., and Takagi, T. (1983). Determination of molecular weight of membrane proteins by the use of low-angle laser light scattering combined with high performance gel chromatography in the presence of a nonionic surfactant. *Biochim. Biophys. Acta. 747*, 291–297.

Maezawa, S. and Takagi, T. (1983). Monitoring of the elution from a high-performance gel chromatography column by a spectrophotometer, a low-angle laser light scattering photometer and a precision differential refractometer as a versatile way to determine protein molecular weight. *J. Chromatogr. 280*, 124–130.

Nakao, T., Ohno, T., Nakao, M., Maeki, G., Tsukita, S., and Ishikawa, H. (1983a). Monomeric and trimeric structures of active Na,K-ATPase in $C_{12}E_8$ solution. *Biochem. Biophys. Res. Commun. 113*, 361–367.

Nakao, T., Ohno-Fujitani, T., and Nakao, M. (1983b). Sodium and potassium ion dependent change in oligomerization of Na,K-ATPase in $C_{12}E_8$ detected by low-angle laser light scattering technique in combination with high performance porous silica gel chromatography, *J. Biochem. 94*, 689–697.

Ouano, A. C. and Kaye, W. (1974). Gel-permeation chromatography: X. Molecular weight detection by low-angle laser light scattering. *J. Polym. Sci. Polym. Chem. Ed. 12*, 1151–1162.

Putzeys, P. and Brosteaux, J. (1935). The scattering of light in protein solutions. *Trans. Faraday Soc. 31*, 1314–1325.

Schmid, K. S. (1975). α_1-Acid glycoprotein. In: *The Plasma Proteins*, Vol. I, F. W. Putnam (Ed.). Academic Press, New York, pp. 183–228.

Schmid, K., Kaufman, H., Isemura, S., Bauer, F., Emura, J., Motoyama, T., Ishiguro, M., and Manno, S. (1973). Structure of α_1-acid glycoprotein. The complete amino acid sequence, multiple amino acid substitution, and homology with the immunoglobulin. *Biochemistry 12*, 2711–2724.

Takagi, T. (1981). Confirmation of molecular weight of *Aspergillus oryzae* α-amylase using the low-angle laser light scattering technique in combination with high pressure silica gel chromatography. *J. Biochem. 89*, 363–368.

Takagi, T. (1982). Assessment study on the use of the low-angle laser light scattering technique in combination with the high performance porous silica gel chromatography. In: *Protides of the Biological Fluids*, Colloquim 30, H. Peeters (Ed.). Pergamon Press, Oxford, pp. 701–704.

Takagi, T., Kameyama, K., and Ui, N. (1981). Assessment study on the high pressure GPC-low angle laser light scattering technique as a method of protein molecular weight determination. *Proceedings of the 34th Symposium of Protein Structure*, pp. 105–108 (in Japanese).

Tanford, C. (1961). Light scattering. In: *Physical Chemistry of Macromolecules*. John Wiley, New York, pp. 275–316.

Progress in HPLC, Vol. 1, pp. 43—57
Parvez *et al.* (Eds)
© 1985 VNU Science Press

Gel permeation of proteins by high-performance gel chromatography in denaturing solvents

KATSUTOSHI KONISHI

Department of Biophysical Chemistry, Dokkyo University School of Medicine, Mibu, Tochigi 321-02, Japan

INTRODUCTION

The field of aqueous gel chromatography is one of rapid progress in both theoretical and experimental aspects.

High-performance gel chromatography (HPGC) for the analytical determinations of biopolymer systems has made rapid advance in the last five years as a result of the appearance of columns packed with rigid matrices TSK-GEL SW made by Toyo Soda, Co., Tokyo.

In 1979, it was demonstrated that the columns can be used effectively to estimate the molecular weights of protein polypeptide chains in the presence of sodium dodecyl sulfate (Imamura *et al.*, 1979) and Ui (1979) has also shown that the columns are useful for the rapid determination of molecular weights of protein polypeptide chains in 6 M guanidine hydrochloride, a random coil-producing solvent. Subsequently the usefulness of these columns in conjunction with high-speed liquid chromatograph has been appreciated by many researchers.

This chapter deals with the gel filtration behavior of a number of protein polypeptide chains in various kinds of denaturants such as SDS, urea, and guanidine hydrochloride and describes the electrostatic interaction between the gel surfaces and protein polypeptide chains.

EXPERIMENTAL CONDITION FOR HPGC

The chromatographic studies on the polypeptides were performed with a Model HLC 802 liquid chromatograph (Toyo Soda) equipped with a TSK-GEL G3000SW or G4000SW column (7.5 × 600 mm) and a precolumn (7.5 × 75 mm). Twenty μl of each sample was charged on the column previously equilibrated with the elution buffer. Ultraviolet absorption with a Schoeffer Spectro Flow monitor SF 770 at 220 or 280 nm was used to detect the samples elution peaks.

The void volume (V_0) and inner volume (V_i) were determined from the elution volume of blue dextran 200 and 2-mercaptoethanol, respectively. The distribution coefficient, K_d, was calculated according to the equation $K_d = (V_e - V_0)/V_i$, where V_e is the elution volume of each native protein or polypeptide.

HPGC OF NATIVE PROTEINS UNDER NONDENATURING CONDITION

The behavior of native proteins in TSK-GEL SW type columns depends on the salt concentration of the elution buffer. Figure 1 shows plots of the values of the distribution coefficient, K_d, for typical native proteins, against the logarithm of sodium phosphate concentration. The salt concentration dependence of K_d values was different between acidic proteins and basic proteins. In acidic proteins such as serum albumin, ovalbumin, β-lactoglobulin, and insulin A chain, the K_d values increased with the rise of the concentration of sodium phosphate. Basic proteins such as chymotrypsinogen A, lysozyme, and cytochrome c are absorbed with the resin. The K_d value of myoglobin was not altered except in extremely low salt concentrations.

It is obvious that low concentrations of sodium phosphate, below 10 mM, are unsuitable for size exclusion. High concentration conditions of sodium phosphate, such as 0.2 M, are required to elute highly basic proteins such as lysozyme and cytochrome c. The elution peak, however, may be broadened in extremely high concentrations (above 0.4 M).

It appears that these phenomena are attributable to the interaction of proteins and negatively charged gel surfaces caused by silanol groups. As described later

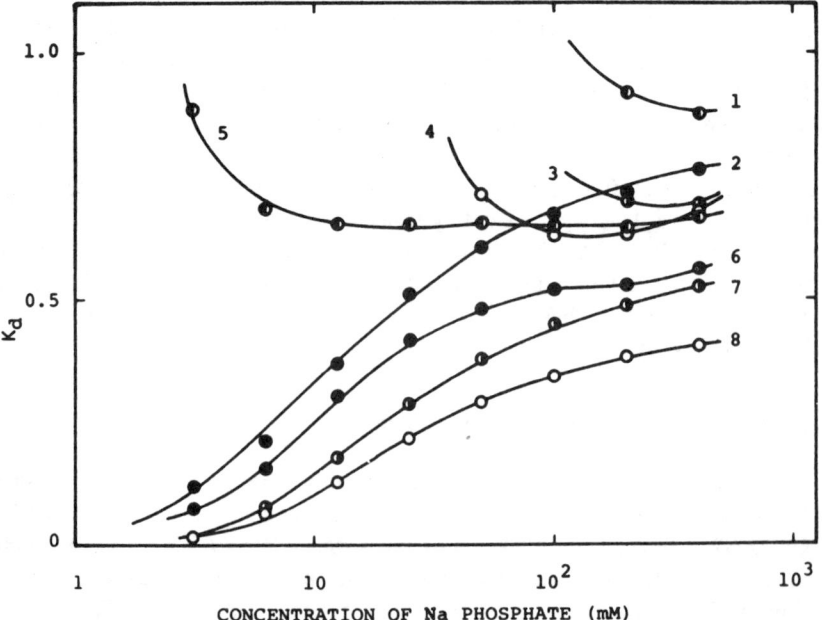

Figure 1. Relationships between distribution coefficient, K_d, of native proteins and sodium phosphate concentration. The proteins are: 1, lysozyme; 2, insulin A chain; 3, cytochrome c; 4, chymotrypsinogen; 5, myoglobin; 6, β-lactoglobulin; 7, ovalbumin; 8, serum albumin.

the diffuse double layers on gel surfaces and proteins will be compressed at high salt concentrations in the elution buffer and acidic proteins may be consequently capable of penetrating the gel interior. It is supposed that the elution positions of basic proteins are retarded by the attractive interaction between positively charged proteins and gel surfaces.

HPGC OF PROTEIN POLYPEPTIDE CHAINS IN DENATURING SOLVENTS

The SDS solvent system
For the experiments some polypeptides were prepared by cyanogen bromide (CNBr) treatment of eleven native proteins that are listed in Table 1. Those samples were also used for the experiments in other denaturing solvents.

Effects of sodium phosphate concentration
The relationship of molecular weights versus K_d of polypeptides was significantly influenced by the concentration of sodium phosphate. Semi-logarithmic plots of the molecular weight and K_d of polypeptides in 0.1% SDS are shown Fig. 2.

Table 1. Cyanogen bromide fragments used in gel filtration experiments

CNBr fragment		Molecular weight
Serum albumin	CNBr I	9 860
	CNBr III	29 520
Aldolase	CNBr I	17 710
	CNBr II	7 370
	CNBr III	2 040
	CNBr IV	11 900
Carboxypeptidase A	CNBr I	2 660
	CNBr II	9 230
	CNBr III	22 120
Pepsin	CNBr I	8 630
	CNBr II	12 640
Chymotrypsinogen A	CNBr I	19 180
	CNBr II	1 060
	CNBr III	5 560
Trypsinogen	CNBr I	9 900
α-Chymotrypsin	B chain	13 920
	CNBr I	3 580
	CNBr III	5 560
Trypsin	CNBr I	9 100
Myoglobin	CNBr I	6 230
	CNBr III	2 510
Lysozyme	CNBr I	1 270
	CNBr II	10 160
	CNBr III	2 860
Cytochrome *c*	CNBr I	7 030
	CNBr II	1 780
	CNBr III	2 780

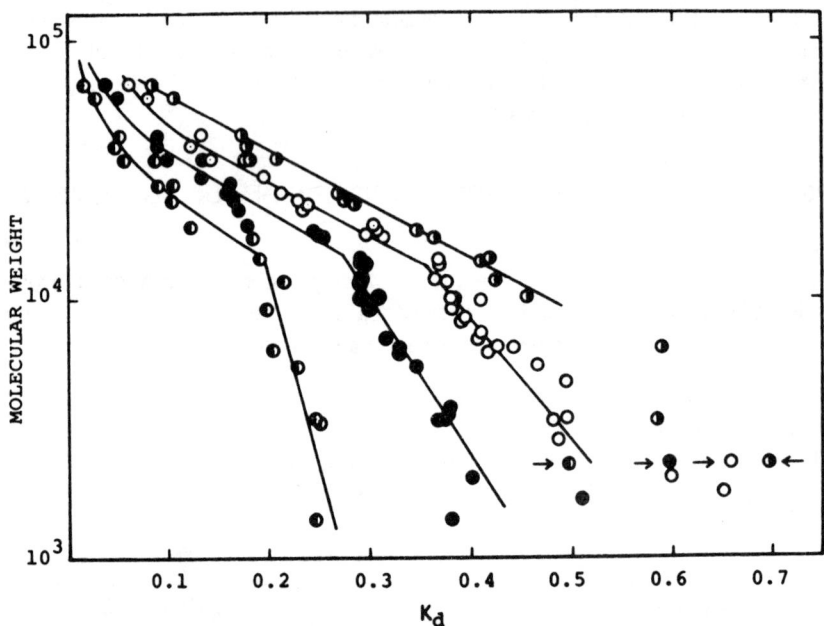

Figure 2. Sodium phosphate concentration dependence of the plots of log molecular weight versus K_d of polypeptides in SDS. Sodium phosphate concentrations in the eluents shown are 0.025 M (◓), 0.05 M (●), 0.1 M (○) and 0.20 M (◑). The solid arrows represent the data for the sample of insulin *A* chain (s-carboxymethylated).

There was a difference in the influence of the salt concentration between poly-peptides having a molecular weight higher than about 15 000 and those lower than about 15 000.

The detailed examination of many different polypeptides showed that the slope of the plot did not change with the variation of the salt concentration, but only the K_d values increased with the rise of the salt concentration. The increments of K_d values were practically suppressed at about 0.2 M sodium phosphate, higher than this its concentration gave broadened peaks.

The slope of the plot for polypeptides having lower molecular weights (below 15 000) was steeper than those for polypeptides in lower salt concentrations. As a result, the plot had an inflection at around 15 000. It is considered that the phenomenon may be based on the difference between the hydrodynamic properties of protein polypeptide chains bound by SDS and those of short-chain polypeptides bound by SDS.

Effects of SDS concentration
The effects of SDS concentration on the exclusion patterns and the positions of peaks were investigated in the concentration range 0.05–1% SDS.

No marked effect of SDS concentration on the plots between log molecular weights and K_d was detected in the region above 10 000 for the molecular weights of polypeptides. For polypeptides having low molecular weights, below 10 000, the plot in 1% SDS lost the linearity and became steeper. In 0.05% SDS, on the other hand, the elution peaks for the polypeptides having lower molecular weights were greatly broadened and became obscure.

We selected empirically the concentration condition of 0.2 M sodium phosphate containing 0.2% SDS for the preparation of the calibration curve.

The relationship between log molecular weights and K_d for 45 different polypeptides is shown in Fig. 3. The plot gives good linearity, particularly for polypeptides and an inflection on the curve that appeared around 15 000 is moderated.

An example for the separation of the mixture of polypeptides in 0.2 M sodium phosphate (pH 7.0) containing 0.2% SDS is shown in Fig. 4. Each peak of serum albumin, pepsin, trypsinogen, myoglobin, cytochrome c, and insulin A chain was completely separated.

Takagi (1981) indicated that by using two columns better resolution of the elution peaks was obtained than using only one.

As is well known, the solution of surfactant such as SDS had a critical micelle concentration (c.m.c.). When the concentration of SDS exceeds the c.m.c. in which the concentration of SDS molecules dispersed becomes constant, surplus SDS molecules are used up to form micelles. The amount of SDS-binding to the polypeptides is saturated at the c.m.c. because only SDS molecules dispersed take part in the binding. As the c.m.c. depends on the salt concentration of buffer solution (Emerson and Holzter, 1965), special attention should be paid to the experimental conditions in SDS system.

As described above, the resolution of the elution peaks of polypeptides having low (below 10 000) molecular weights becomes less according with the increase of the SDS concentration from 0.1% to 1%. This phenomenon is caused by the

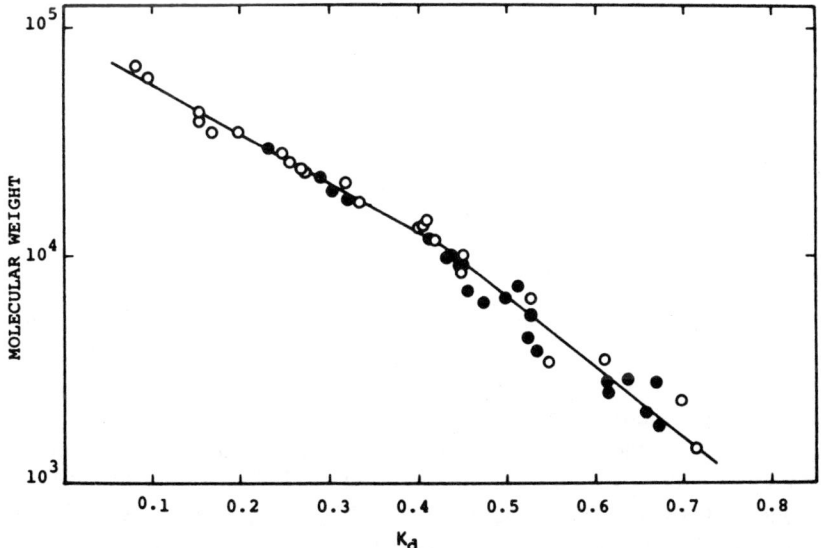

Figure 3. Plot of log molecular weight versus K_d of 45 polypeptides in 0.2% SDS and 0.2 M sodium phosphate buffer (pH 7.0). Solid circles represent the data for CNBr fragments.

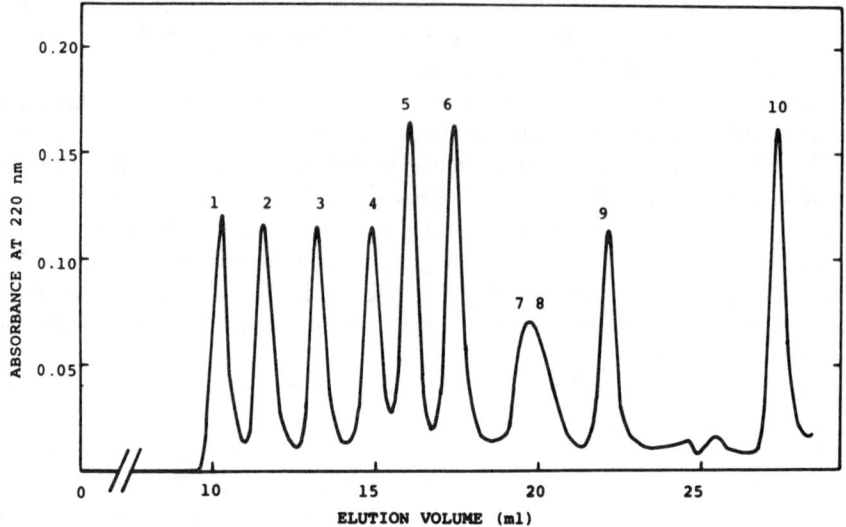

Figure 4. An elution profile of a mixture of polypeptides from a G3000SW column. The elution was carried out in 0.2% SDS and 0.2 M sodium phosphate buffer (pH 7.0). Peaks are: 1, blue dextran; 2, serum albumin; 3, pepsin; 4, trypsinogen; 5, myoglobin; 6, cytochrome *c*; 7, aprotinin; 8, insulin *B* chain; 9, insulin *A* chain; 10, 2-mercaptoethanol.

reason why the behavior of SDS-oligopeptide chain complexes depends on the concentration of free SDS molecules and effects upon the separation of the short chain polypeptides in the chromatography.

6 M Guanidine hydrochloride system
The first attempt to perform high-performance gel chromatography in 6 M guanidine hydrochloride by use of TSK-GEL G3000SW and G4000SW was made by Ui (1979).

As reduced simple proteins in concentrated guanidine hydrochloride solutions behave hydrodynamically as randomly coiled linear homopolymers whose radius of gyration is a simple function of the molecular weight, a linear relationship is expected between $M^{0.555}$ and $K_d^{1/3}$ (Ui, 1979).

Figure 5 shows that the plot of $M^{0.555}$ versus $K_d^{1/3}$ has good linearity, where M is the molecular weight and K_d is the distribution coefficients. Ui indicated that the accuracy and reproducibility of the elution volumes measured and, hence, of the calculated K_d values, together with the linearity of the plots of $M^{0.555}$ versus $K_d^{1/3}$, produced the basis for reliable, as well as convenient, estimation of the molecular weights of polypeptides.

Ui has applied the method to study the hydrodynamic behavior of reduced glycopolypeptides (given in Table 2) with TSK-GEL G3000SW (Ui, 1981). The relationship between $M^{0.555}$ and $K_d^{1/3}$ for the glycopolypeptides is shown in Fig. 6. According to his study, this method appears to be highly useful for the rapid estimation of the molecular weight not only of simple proteins but also of glycopolypeptides, even though two carbohydrate-rich glycopolypeptides such

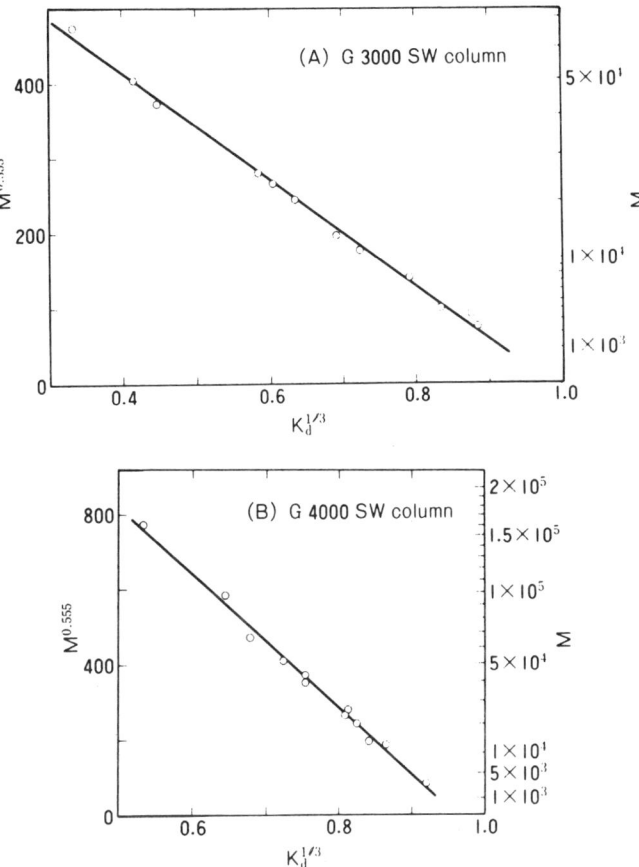

Figure 5. Plots of $M^{0.555}$ versus $K_d^{1/3}$ of various polypeptides in 6 M guanidine hydrochloride with G3000SW (A) and G4000SW (B) columns. Data from Ui (1979).

as α-acid glycoprotein and acid carboxypeptidase, with 41.3% and 21.6% carbohydrate contents, respectively, give underestimated molecular weights.

Ui also suggested that the absence of a correlation between the extent of the deviation of molecular weights given in Table 2 and carbohydrated content might indicate that the number, location, length, and structure of oligosaccharide branches have different influences in determining the overall hydrodynamic dimension of glycopolypeptides in 6 M guanidine hydrochloride.

Besides the orthodox method to relate K_d to M as discussed by Ui, we show the relationship between log molecular weight and K_d of protein polypeptides in 6 M guanidine hydrochloride and 0.05 M sodium phosphate buffer (pH 7.0) in Fig. 7. The linearity of the plot was maintained in the molecular weight range of 2000 to 70 000. The K_d values of polypeptides in 6 M guanidine hydrochloride were scarcely affected by the change of sodium phosphate concentration.

This fact indicates that a high concentration of electrolytes, such as concentrated guanidine hydrochloride, may effectively eliminate the electrostatic interactions

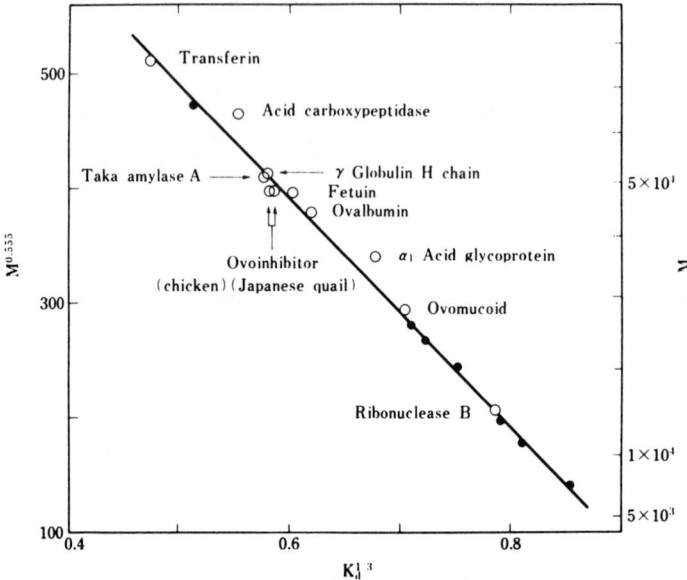

Figure 6. Relationship between $M^{0.555}$ and $K_d^{1/3}$ obtained in 6 M guanidine hydrochloride. The straight line was drawn using the data for simple polypeptides (●) and compared with the data for glycopolypeptides (○). Data from Ui (1981).

Table 2. Distribution coefficient, K_d, and molecular weight, M, of glycopeptides. Data from Ui (1981)

Protein	M	CHO (%)	K_d	M from K_d	Difference in M values (%)
Transferin	76 000	6	0.107 ± 0.003	77 700	+2.2
Acid carboxypeptidase (*A. niger*)	64 000	21.6	0.170 ± 0.002	57 600	−10.0
γ-Globulin H chain	51 500	4.9	0.195 ± 0.002	51 400	−0.2
Taka-amylase A	51 000	2.7	0.192 ± 0.003	52 100	+2.2
Ovoinhibitor (chicken)	48 300	12	0.197 ± 0.004	51 000	+5.6
Ovoinhibitor (Japanese quail)	48 300	12	0.202 ± 0.003	50 000	+3.5
Fetuin	48 000	22	0.220 ± 0.005	46 300	−3.5
Ovalbumin	44 300	3.6	0.238 ± 0.005	42 800	−3.4
α_1-Acid glycoprotein	36 500	41.3	0.312 ± 0.004	31 500	−13.7
Ovomucoid	28 000	23	0.350 ± 0.005	26 900	−3.9
Ribonuclease B	14 900	8.1	0.487 ± 0.006	14 700	−1.3

between polypeptides and gel surfaces. In the concentrations above 0.4 M of sodium phosphate, however, good results will not be expected for the same reason as it was observed for native proteins and polypeptide-SDS complexes.

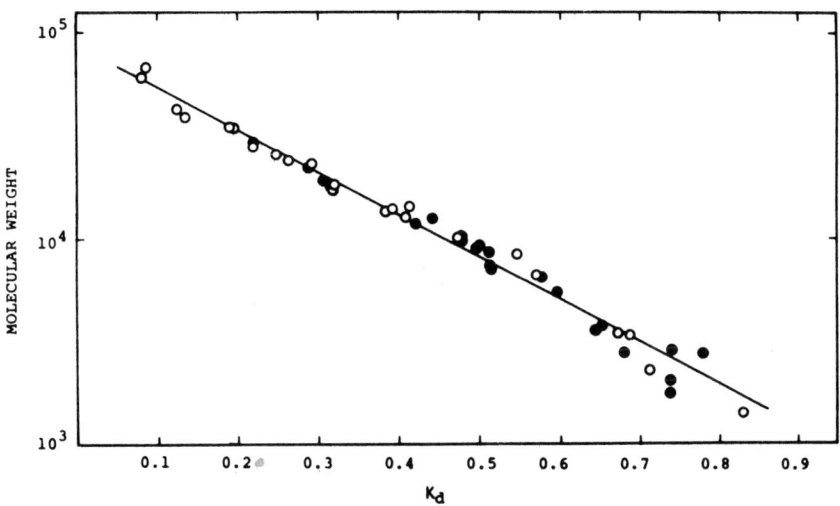

Figure 7. Relationship between log molecular weights and K_d for 45 polypeptides in 6 M guanidine hydrochloride and 0.05 M sodium phosphate buffer (pH 7.0). Solid circles represent the data for CNBr fragments.

8 M Urea system

The polypeptides in 8 M urea were also similar in chromatographic behavior to native proteins but the effect of sodium phosphate concentration was not as drastic as those to native proteins.

Figure 8 shows the relationship between log molecular weight and K_d of

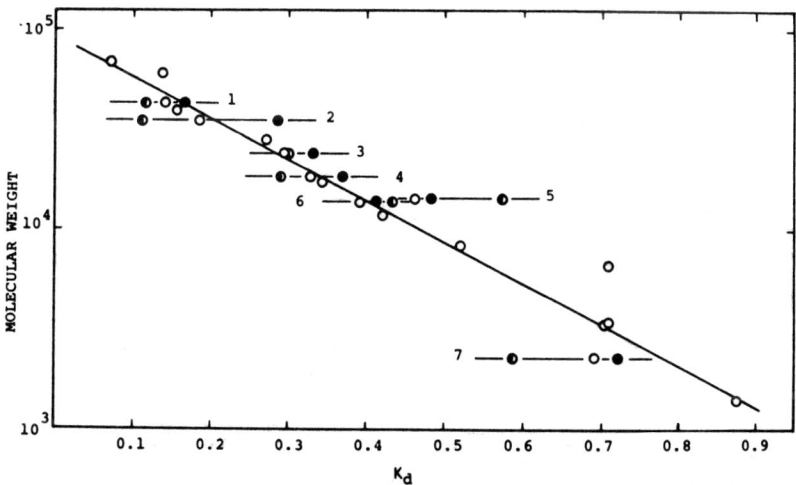

Figure 8. Relationship between log molecular weights and K_d for polypeptide 8 M urea at various concentrations of sodium phosphate buffer (pH 7.0). Proteins are: 1, ovalbumin; 2, pepsin; 3, trypsinogen; 4, β-lactoglobulin; 5, lysozyme; 6, ribonuclease; 7, insulin A chain. The concentrations of sodium phosphate buffer (pH 7.0) in the eluents were 0.05 M (◐), 0.20 M (O), and 0.4 M (●).

polypeptides reduced on disulfate bonds in 8 M urea. A relatively good linearity is achieves in the elution using 0.2 M sodium phosphate. The plots of both highly acidic and basic polypeptides deviated from the line in lower salt concentrations. At higher sodium phosphate concentrations, on the other hand, K_d values were larger than those evaluated on the basis of the electric properties of polypeptides.

Systematical consideration for chromatographic behavior in denaturing solvents
In order to demonstrate the chromatographic behavior of native proteins and polypeptides in various kinds of denaturing solvents, the TSK-GEL G3000SW or G4000SW column must be calibrated by the method based on K_d values and Stokes radii of proteins and polypeptides. We adopted the method proposed by Ackers (1967); Fish *et al.* (1970); and Freytag *et al.* (1979), for this purpose. The relationship between Stokes radii, R_s, calculated from intrinsic viscosity data and erf^{-1} $(1 - K_d)$ of some proteins and polypeptides in various denaturing solvents is shown in Fig. 9.

The plot for some native proteins and polypeptides in 6 M guanidine chloride and 8 M urea followed a common curve except for native lysozyme. The deviation of the protein may be due to the attraction of positively charged protein with negatively charged gel surfaces.

The plot of some polypeptide-SDS complexes which had highly negative charges also substantially followed the common curve. The repulsive interaction between negatively charged polypeptide-SDS complexes and gel surfaces may be taken as negligibly small in the selected condition of buffer solution. These results suggest

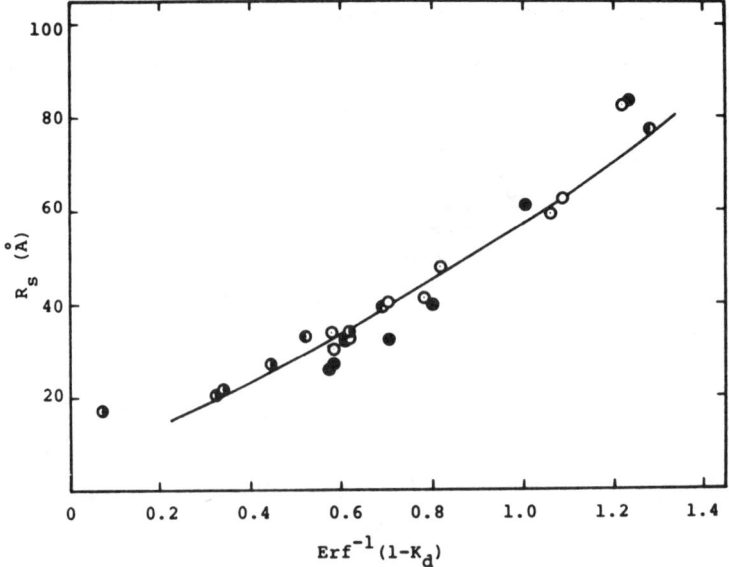

Figure 9. Relationship between R_s and erf^{-1} $(1 - K_d)$ of some polypeptides in various eluents. The eluents were: 0.20 M sodium phosphate (◗), 6 M guanidine hydrochloride and 0.05 M sodium phosphate (○); 8 M urea and 0.20 M sodium phosphate (◐); 0.2% SDS and 0.20 M sodium phosphate (●).

that proteins and polypeptides in denaturing solvents behave like homopolymers. Therefore, these methods can be applied to estimate the molecular weights of proteins and polypeptides.

An application of SDS system to estimate molecular weights of membrane proteins
We applied the SDS system to estimate the molecular weights of acetylcholine receptor subunits, so-called α, β, γ, and δ, from *Narke japonica* (Imamura *et al.*, 1983). Four kinds of subunit samples were prepared according to the method described by Haggerty and Froehner (1981). High performance gel chromatography for those subunit samples was carried out by a Toyo Soda HLC-802 high-speed liquid chromatograph equipped with a TSK-GEL G4000SW (7.5 × 600 mm) in 0.2% SDS and 0.2 M sodium phosphate buffer (pH 7.0). The distribution coefficient, K_d, was calculated in accordance with the method already given.

As shown in Fig. 10, the relationship between K_d and the molecular weights of standard protein showed good linearity. The values of K_d for the α-, β-, γ-, and δ- subunits gave molecular weights of 36 000, 43 000, 45 000, and 48 000, respectively. These values agree with the molecular weights determined by means of the low angle laser light scattering technique in combination with high-performance gel chromatography (HPGC-LALLS), introduced by Takagi in Chapter 3 of this book. But the values were smaller than those obtained by SDS-polyacrylamide gel electrophoresis (Imamura *et al.*, 1983).

GENERAL CONSIDERATION FOR HPGC

TSK-GEL SW series gel packings are particulate silica matrices covalently bonded with hydrophilic compounds (Fukano *et al.*, 1978). The gel surfaces have negative

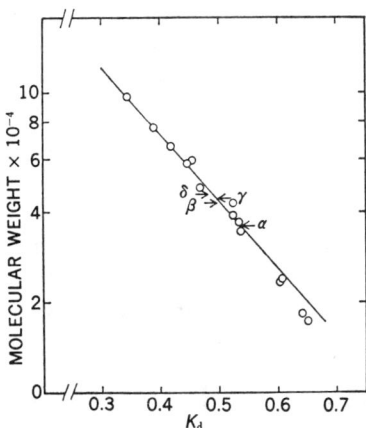

Figure 10. Relationship between log molecular weights and K_d for standard proteins with G4000SW column in 0.2% SDS and 0.2 M sodium phosphate buffer (pH 7.0). Proteins are phosphorylase *a*, transferrin, BSA, γ-globulin H and L chains, catalase, pyruvate kinase, ovalbumin, aldolase, pepsin, alcohol dehydrogenase, trypsinogen, β-lactoglobulin, and myoglobin. Arrows indicate the values for the Ach R subunits.

charges which are probably due to traces of silanol groups remaining unreacted and consequently affected by the cation concentration in the elution buffers (Imamura et al., 1981). It must be considered for the estimation of the hydrodynamic properties such as Stokes radii of polypeptides to adopt the values of the distribution coefficient, K_d, which is obtained by extrapolation to infinitely high concentrations of cations. The zero-dimension model for gel filtration presented by Rodbard and Chrambach (1970) and the electric double layer model by Gouy-Chapman (Verwey and Overbeek, 1948) were taken for this purpose.

The zero-dimensional model gave the following equation for the relationship between distribution coefficient, K_d, and radius, R of protein particle

$$K_d = \exp\{-\tfrac{4}{3}n\pi(R + r)^3\} \tag{1}$$

where n and r are the number of gel particles per unit volume and the radius of gel particle, respectively. Moreover K_d is related to the differently distribution coefficient, K_{av}, which was defined by Laurent and Killander (1964) with respect to the total bed volume of the packed chromatography column, V_t. This is given by the relation (Fish, 1975)

$$K_{av} = K_d \frac{V_i}{V_i + V_g} \tag{2}$$

where V_i is inner volume and V_g represents to the total volume of the packed column which is occupied by the matrix of the chromatographly support medium.

When the surfaces of the gel and protein particle have electric charges, the thickness of the diffuse double layers on both surfaces are influenced by the concentration of counter ion in the medium. If we assume that those surfaces are approximately flat and wide and that the electrical potentials on the surfaces, ψ_0, are low, the following equation can be introduced from the Gouy-Chapman theory as described in the textbook on surface chemistry by, Verwey and Overbeek (1948)

$$\psi = \psi_0 \exp(-\chi x), \qquad \chi = \{(8\pi n_0 Z^2 e^2)/(\epsilon k T)\}^{1/2} \tag{3}$$

where ψ is the electrical potential at a distance, x, from the surface of the gel or protein particle, χ is the reciprocal of the 'thickness' of the diffuse layer, and n_0 is the number of counter ions per cubic meter at a distance far from those surfaces. When a charged protein particle having the same sign as the surface of the gel particle can approach the surface to the distance where the electrical potentials on the surfaces of the gel and protein particle decrease to ψ, Equation (1) can be expressed as follows:

$$K_{av, app} = \exp\{-\tfrac{4}{3}n\pi(R + r + x + y)^3\} \tag{4}$$

where x and y are the distances from the gel and protein particle, respectively, and $K_{av, app}$ is the apparent value of K_{av}. On substituting Equation (3) into (4), the following equation is derived

$$\{\ln(1/K_{av,app})\}^{1/3}$$

$$= (\tfrac{4}{3}n\pi)^{1/3}[R + r + \{(\epsilon k T)/(8\pi e^2 Z^2)\}^{1/2} n_0^{-1/2} x \ln(\psi_0 \psi_0'/\psi^2)]. \tag{5}$$

where ψ_0' is the surface potential of the protein particle. If we assume that ψ_0, ψ_0', R, and r are not greatly influenced by the concentration of counter ion in the medium, the following equation is deduced

$$(\ln K_{av, app}^{-1})^{1/3} = A + BC^{-1/2} \qquad (6)$$

where C is the molar concentration of counter ion in the medium. The constants A and B are related to the radius and surface potential of the protein particle, respectively.

Figure 11 shows the relationship between $(\ln K_{av, app}^{-1})$ and $C^{-1/2}$ for various kinds of simple proteins. The plots gave good linearities for most of the proteins except for the basic one. The slopes of the lines differed depending on the kinds of proteins. This indicates that Equation (6) is approximately valid for many proteins. The equation, however, cannot be applied to highly basic proteins because of the attractive interactions between gel surfaces and these proteins. The constant B in Equation (6) may depend on the electrical potential of protein surface. There are obvious correlations between the B values obtained from the slope of the line in Fig. 11 and the isoelectric points of various proteins listed in Table 3. It is suggested that the electrical properties of protein surfaces should be predictable from the result of the measurements for the cation concentration of K_d.

The values of $(\ln K_{av}^{-1})^{1/3}$ can be obtained by extrapolation of the linear $(\ln K_{av, app}^{-1})^{1/3}$ and Stokes radii of standard proteins calculated from the intrinsic

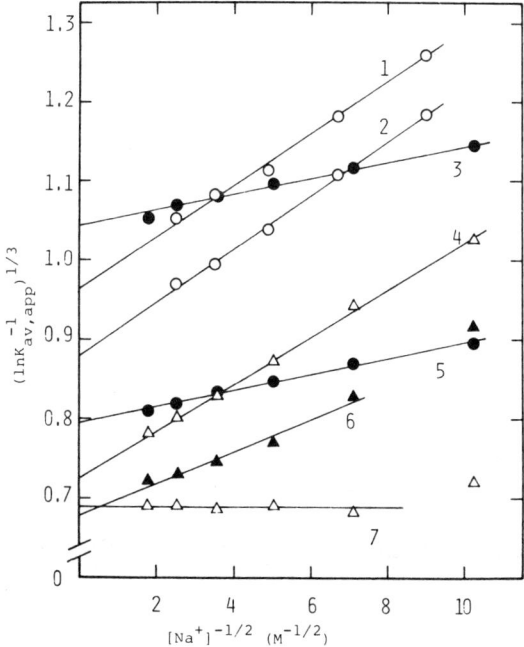

Figure 11. Relationships between $(\ln K_{av, app}^{-1})^{1/3}$ and $[Na^+]^{-1/2}$ for proteins. Lines are: 1, Ach R dimer; 2, Ach R monomer; 3, fibrinogen; 4, BSA; 5, aldolase; 6, β-lactoglobulin; 7. hemoglobin.

K. Konishi

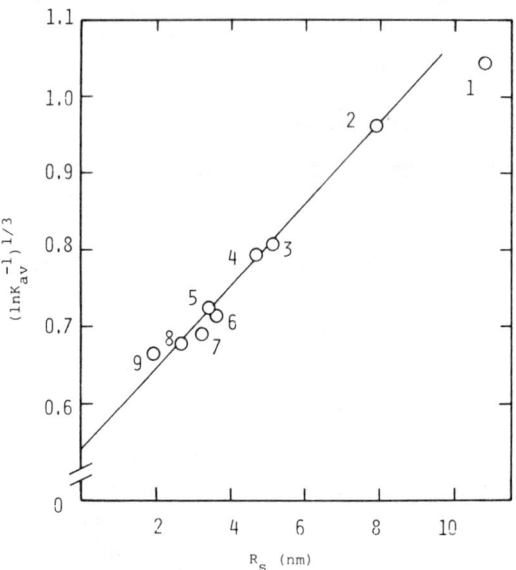

Figure 12. Relationship between $(\ln K_{av}^{-1})^{1/3}$ and Stokes radii of proteins. The values of $(\ln K_{av}^{-1})^{1/3}$ were obtained by extrapolation of the linear $(\ln K_{av,app}^{-1})^{1/3}$ versus $[Na^+]^{1/2}$ plots to the intercepts. Proteins are: 1, fibrinogen; 2, thyroglobulin; 3, γ-globulin; 4, aldolase; 5, BSA; 6, transferrin; 7, hemoglobin; 8, β-lactoglobulin; 9, myoglobin. Arrows indicate the values for the Ach R dimer and monomer.

Table 3. Comparison of B values and isoelectric points of proteins

Proteins	Isoelectric points	B values $(M^{1/2} \times 10^2)$
Thyroglobulin	4.5	3.0
BSA	4.7–4.9	3.0
β-Lactoglobulin	5.1	2.0
Transferrin	5.2	2.6
Fibrinogen	5.5–5.8	1.0
γ-Globulin	5.8–7.3	0.8
Hemoglobin	6.8–7.0	0.0
Myoglobin	8.1–8.2	0.0
Aldolase	8.2–8.6	1.0
AChR dimer monomer	4.9	3.3

viscosity data (Fish, 1975) which are plotted in Fig. 12. Good linearity was achieved as expected.

CONCLUSIONS

High-performance gel chromatography (HPGC) using the TSK-GEL SW series gel packings is an instrument for the quantitative separation and analysis of polypeptides in denaturing solvents because of the excellent resolution and reproducibility of the gels. HPGC in denaturing solvents appears to be the best method for estimating the molecular weights of simple polypeptides and glyco-polypeptides rapidly and conveniently. The method may be especially useful to estimate the molecular weights of membrane proteins rapidly. In addition, not only the hydrodynamic properties of polypeptides but also those electrical properties can be predicted from the detailed measurements of their behavior in HPGC.

REFERENCES

Ackers, G. K. (1967). A new calibulation procedure for gel filtration columns. *J. Biol. Chem.* *242*, 3237–3238.

Emerson, M. F. and Holtzer, A. (1965). On the ionic strength dependence of micelle number. *J. Phys. Chem.* *69*, 3718–3721.

Fish, W. W., Reynolds, J. A., and Tanford, C. (1970). Gel chromatography of proteins in denaturing solvents. *J. Biol. Chem.* *245*, 5166–5168.

Fish, W. W. (1975). Determination of the molecular weights of membrane proteins and poly-peptides. In: *Methods in Membrane Biology*, Vol. 4, Biophysical approaches, E. D. Korn (Eds.). Plenum Press, New York, pp. 189–276.

Freytag, J. W., Noelken, M. E., and Hudson, B. G. (1979). Physical properties of collagen sodium dodecyl sulfate complexes. *Biochemistry* *18*, 4761–4768.

Fukano, K., Komiya, K., Sasaki, H., and Hashimoto, T. (1978). Evaluation of new supports for high-pressure aqueous gel permeation chromatography. *J. Chromatogr.* *166*, 47–54.

Haggerty, J. G. and Froehner (1981). Restoration of ^{125}I-α-bungarotoxin binding activity to the α subunit of *Torpedo* acetylcholine receptor isolated by gel electrophoresis in sodium dodecyl sulfate. *J. Biol. Chem.* *256*, 8294–8297.

Imamura, T., Konishi, K., Yokoyama, M., and Konishi K. (1979). High-speed gel filtration of polypeptides in sodium dodecyl sulfate. *J. Biochem.* *86*, 639–642.

Imanura, T., Konishi. K., Yokoyama, M., and Konishi, K. (1981). High-speed gel filtration of polypeptides in some denaturants. *J. Liq. Chromatogr.* *4*, 613–627.

Imamura, T., Konishi, K., and Konishi, K. (1982). Physico-chemical properties of acetylcholine receptor proteins from *Narke japonica. J. Biochem.* *92*, 1901–1910.

Imamura, T., Konishi, K., and Konishi, K. (1983). Molecular weights of acetylcholine receptor subunits from *Narke japonica. J. Biochem.* *94*, 917–923.

Laurent, T. C. and Killander, J. (1964). A theory of gel filtration and its experimental verifica-tion. *J. Chromatgr.* *14*, 317–330.

Rodbard, D. and Chrambach, A. (1970). Unified theory for gel electrophoresis and gel filtration. *J. Biochem.* *65*, 970–977.

Takagi, T. (1981). High-performance liquid chromatography of protein polypeptides on porous silica gel columns (TSK–GEL SW) in the presense of sodium dodecyl sulphate: comparison with SDS-polyacrylamide gel electrophoresis. *J. Chromatogr.* *219*, 123–127.

Ui, N. (1979). Rapid estimation of the molecular weights of protein polypeptide chains using high-pressure liquid chromatography in 6 M guanidine hydrochloride. *Anal. Biochem.* *97*, 65–71.

Ui, N. (1981). High-speed gel filtration of glycopolypeptides in 6 M guanidine hydrochloride. *J. Chromatogr.* *215*, 289–294.

Verwey, E. J. W. and Overbeek, J. Th. G. (1948). Distribution of the electric charge and potential in the electro-chemical double layer. In: *Theory of the Stability of Lyophobic Colloids*. Elsevier, Amsterdam, pp. 22–37.

Progress in HPLC, Vol. 1, pp. 59—81
Parvez *et al.* (Eds)
© 1985 VNU Science Press

Multiple forms of cytochrome P-450 separable by high-performance liquid chromatography

YOSHIHIKO FUNAE,[1]* ALVIN N. KOTAKE,[3] and KENJIRO YAMAMOTO[2]

[1] Laboratory of Chemistry and [2] Department of Pharmacology, Osaka City University Medical School, Abeno-ku, Osaka 545 Japan and [3] Committee on Clinical Pharmacology, Department of Pharmacological and Physiological Science, University of Chicago, Chicago, IL 60637, USA

INTRODUCTION

Macroporous support materials coated covalently with a layer of glycerylpropyl silyl groups have been used in HPLC. Ion-exchange support for HPLC has been used to separate plasma proteins (Chang *et al.*, 1976; Funae *et al.*, 1982), iso-enzymes of lactic dehydrogenase (Schlabach *et al.*, 1978; Vacik and Toren, 1982), urinary kallikrein (Funae *et al.*, 1983), and hemoglobin (Gooding *et al.*, 1979). We have succeeded in separating a membrane-bound protein, cytochrome P-450 (Kotake and Funae, 1980; Kusunose *et al.*, 1981; Funae and Kotake, 1982; Kohli *et al.*, 1982). Cytochrome P-450 is the terminal oxidase of the hepatic microsomal mixed function oxidase system which plays a central role in the metabolism of steroids, fatty acids, and various exogenous drugs. Many of the major differences in the metabolism of various substrates are due to the presence of different forms of cytochrome P-450 (Conney, 1967; Guengerich, 1979). The presence of multiple forms of cytochrome P-450 is supported by the selective induction of different cytochrome P-450 by various compounds (Sladek and Mannering, 1966; Lu *et al.*, 1972; Ryan *et al.*, 1979). In addition, many groups succeeded in separating, purifying, and identifying multiple forms of cytochrome P-450 from hepatic microsomes, using electrophoresis (Haugen *et al.*, 1975; Ingelman-Sandberg and Gustafsson, 1977; Guengerich, 1978), ion-exchange chromatography (Ryan *et al.*, 1975; Haugen and Coon, 1976; Warner *et al.*, 1978; West *et al.*, 1979; Kohli *et al.*, 1981; Waxman and Walsh, 1982), and immunological studies (Welton *et al.*, 1975; Thomas *et al.*, 1976; Reik *et al.*, 1982). The number of forms of cytochrome P-450 in normal rat or rabbit hepatic microsomes and the qualitative and quantitative

---

* To whom correspondence should be addressed.

Abbreviations: HPLC, high-performance liquid chromatography; PB, phenobarbital; 3-MC, 3-methylchlanthrene; PCB, polychlorinated biphenyl; SDS—PAGE, sodium dodecyl sulfate polyacrylamide gel electrophoresis.

changes of cytochrome P-450 in the treatment of inducers have not been entirely elucidated. Electrophoresis is not suitable for the quantitation or preparation of purified forms of cytochrome P-450. Ion-exchange chromatography is the most commonly used technique for isolating and characterizing different forms of cytochrome P-450. However, this technique is time consuming and the resolution and reproducibility of the chromatograms leave much to be desired.

We found that HPLC using anion-exchange resin enables separation of multiple forms of rat liver microsomal cytochrome P-450. The chromatographic profiles and the biochemical properties of hepatic and renal cytochrome P-450 induced by PB and 3-MC were elucidated.

MATERIALS AND METHODS

Column support and apparatus
SynChropak AX-300 (also known as Anpac AX-300) anion-exchange resins was obtained from Anspec Co. (Warrenville, Ill. USA). Constametric A and B pumps (Altex Model 100, Berkeley, Calif., USA) with a solvent programmer (Alex Model 400) and dynamic mixer were used for the gradient elution. A sample injector (Rheodyne, Berkeley, CA, USA) with a 400 μl loop was used for injections and spectrophotometers (Hitachi Model 100-20, Tokyo, Japan and Toyo Soda Model UV-8, Tokyo, Japan) with 8 μl of spectrophotometer flow cell were used for detection of hemoprotein.

Column packing
An analytical column (4.6 × 250 mm) was packed with a SynChropak AX-300 resin using a micrometoric Model 705 packer (Norcross, Ga., USA). Three grams of support material were suspended in the packing chamber, in propanol, and the column was packed by pumping water at a rate of 5 ml/min until the pressure stabilized at 4500 psi. The pressure was maintained for 10 min to ensure maximal packing. A 2 μm frit was placed at the inlet.

Preparation of microsomes
Phenobarbital (PB) (40 mg/kg, dissolved in 0.9% NaCl) was administered intra-peritoneally to male Sprague-Dawley rats weighing 180–220 g, daily for 4 days. 3-Methylcholanthrene (3-MC) and polychlorinated biphenyl (PCB) (40 and 80 mg/kg, respectively, dissolved in corn-oil) were given in a single i.p. administration for 3 days. White male rabbits weighing 2.5–3 kg were similarly treated with PB and 3-MC in a dose of 60 and 30 mg/kg in corn-oil for 3 days, respectively. The rats were decapitated and the liver was thoroughly perfused by cannulation into inferior vena cava, *in situ*, with ice-cold 1.15% KCl to avoid contaminating the hemoglobin with microsomes. The liver was homogenized with 3-fold volume of 1.15% KCl using a Potter glass homogenizer. The homogenate was centrifuged at 10 000 g for 30 min and the precipitate was discarded. The supernatant was centrifuged at 100 000 g for 1 h and the resulting microsomal fraction was re-suspended in ice-cold 1.15% KCl and recentrifuged at 100 000 g for 1 h. The microsomal pellet was suspended with 0.1 M potassium phosphate buffer, pH 7.4,

containing 1 mM EDTA, 1 mM dithiothreitol (DTT) and 30% glycerol to the concentration of 10–20 mg protein/ml and stored at −20°C. The rabbit hepatic and renal microsomes were prepared using the method of Kusunose *et al.* (1981).

The specific contents of hepatic cytochrome P-450 of nontreated, PB, 3-MC, and PCB-induced rats were 0.72, 1.44, 1.37, and 1.60 nmol/mg of protein, respectively.

Solubilization of microsomes
Microsomal suspension (4 ml) was solubilized with the addition of 10% sodium cholate (0.5 ml) recrystalized from ethanol-water and 10% Emulgen 913 (Kao-Atlas, Tokyo, Japan) (0.5 ml) at 4°C. When the solubilized solution was not clear, centrifugation at 100 000 g for 30 min was necessary prior to injection into the HPLC. Solubilized microsomes were kept −20°C.

Ion-exchange HPLC of solubilized microsomes
Two Altex pumps equipped with a Model 400 solvent programmer and a Hitachi 100-20 spectrophotometer fitted with an 8 μl flow cell were used. The separations were accomplished using a 4.6 X 250 mm stainless steel column packed with 10 μm Synchropak AX-300 ion-exchange resin. The elution of heme-containing proteins was monitored at 417 nm. Ion-exchange chromatography was carried out at a flow rate of 1.6 ml/min with a linear salt-gradient made by controlling buffer A (0.02 M Tris-acetate, pH 7.5 containing 20% glycerol and 0.2% Emulgen 911) and buffer B (1.0 M sodium acetate was added to buffer A, pH 7.5) using a solvent programmer at 20–25°C. The chromatograms were developed with a 30 min reading for the initial 16–17 min (B% 55–60%) followed by a 3-min gradient. The 3-min gradient was maintained until the elution of cytochrome b_5 was completed. On completion of the chromatogram, a 3-min reverse gradient was initiated and the column was equilibrated with buffer A for 15–20 min.

Methods of analysis
The concentration of cytochrome P-450 or P-448 was determined using the method of Omura and Sato (1964) from the carbon monoxide difference spectrum of dithionite reduced samples, using an extinction coefficient of 91 mM^{-1} cm^{-1}. Assignment of each peak of cytochrome P-450 or cytochrome b_5 was accomplished by spectrophotometrically analyzing the HPLC eluents collected either with a fraction collector or manually. The absolute spectrum of cytochrome P-450 was measured using buffer A as a reference. NADPH cytochrome c reductase activity was determined according to the method of Philips and Langdon (1962).

Measurement of protein concentration
For the measurement of the protein concentration of the eluate on HPLC, we used the following modified method of Lowry *et al.* (1951), because the eluate contained Emulgen, glycerol, and Tris-base, which interfere with the Lowry method. We added SDS (final 1.5%) into the 2% Na$_2$CO$_3$ dissolved in 0.1 N NaOH solution. Bovine serum albumin dissolved in buffer A containing 0.2% Emulgen and 20% glycerol was used as a standard in the modified method. To measure the specific content of cytochrome P-450, 0.5 ml of eluate was used for the assay of protein, without further treatment.

SDS–PAGE

For the preparation of samples, 300 μl of solubilized microsome was injected into the HPLC. Peak fractions containing 30–50 μg of protein were dialyzed against 0.01 M sodium phosphate buffer (pH 7.5) containing 0.1 mM EDTA for 24 h at 4°C. After freeze-drying, the preparations were dissolved in 100 μl of water. The resulting samples were treated with sodium dodecyl sulfate and 2-mercaptoethanol and electrophoresis was performed using Laemmli's (1970) method. The separating gel contained 9% acrylamide and the stacking gel contained 3% acrylamide. The gel was 1 mm thick and was fixed for 60 min in trichloroacetic acid/sulfosalicylic acid/methanol/water, 57.0 g/17.0 g/150 ml/350 ml, stained for 2 h in 0.25% Coomassie Brilliant Blue R-250 in 50% methanol:acetic acid, 9:1 and destained overnight in methanol:acetic acid:water, 10:10:80. The destained gels were scanned by a soft laser densitometer (LKB Model-2202) and the purity of bands was calculated using a data processor (Shimadzu C-RIA, Tokyo, Japan).

RESULTS

The chromatographic profile of cytochrome P-450 of solubilized hepatic microsome obtained from PB pretreated rats is shown in Fig. 1. One to two milligrams

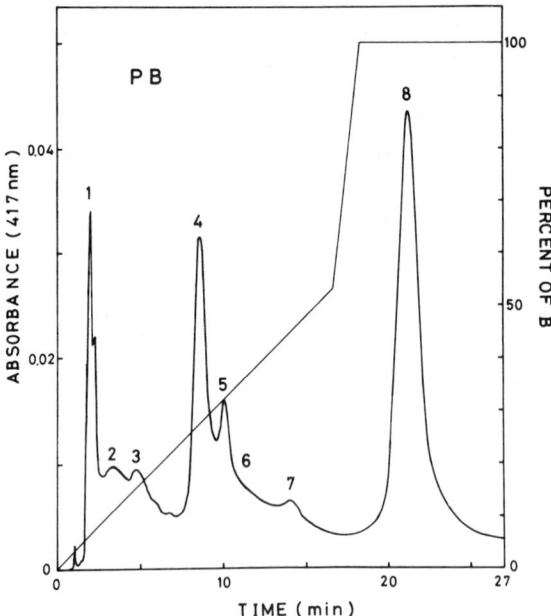

Figure 1. Anion exchange chromatography of phenobarbital-induced rat liver microsome. Analytical column, 4.6 × 250 mm i.d. stainless steel; guard column, 4.1 × 30 mm i.d.; support material, Anpak AX-300 (300 Å pore diameter, 10 μm particle size); temperature, 25°C. Buffer A; 0.02 M Tris-acetate, pH 7.2, containing 0.2% Emulgen 911 and 20% glycerol. Buffer B; 0.8 M sodium acetate is added to buffer A, pH 7.5. Flow rate, 1.6 ml/min; initial linear gradient rate, 100%/30 min; initial pressure, 2200 psi; sample volume, 100 μl (contains 1 mg protein); detector, 417 nm. Percentage of B indicates the concentration of sodium acetate. Time (min) indicates retention time of eluate.

of solubilized microsomal protein was applied to the column and the heme containing peaks were collected, at room temperature. Major peaks (numbers 1, 2 + 3, 4, 5, 7) were found to contain only cytochrome P-450 by the CO-reduced spectrum. Peak 8 was found to contain cytochrome b_5 and was free of cytochrome P-450, as measured spectrophotometrically (Fig. 2). NADPH cytochrome c reductase is eluted at the same position with cytochrome b_5 (Kotake and Funae, 1980). Recovery of cytochrome P-450, as determined by reduced CO difference spectra, was greater than 80% and better than 90% when minor fractions were included. Conversion of cytochrome P-450 to P-420 rarely occurred. HPLC profiles of cytochrome b_5 obtained from nontreated, PB and 3-MC pretreated rats were identical and the cytochrome b_5 fraction did not contain cytochrome P-450.

Conditions of mobile phase
The chromatogram of solubilized microsomes prepared from PB-treated rats using a phosphate buffer instead of a Tris-acetate buffer as a mobile phase is shown in Fig. 3. Peaks 4 and 5 induced by PB treatment were not resolved and were broader than those observed on the chromatograms, using tris-acetate as a mobile phase (Fig. 1). All components eluted slightly earlier than observed in the case of Tris-acetate buffer. Tris-acetate buffer was more efficient than phosphate buffer for the separation of cytochrome P-450.

The effect of sodium chloride gradient on the chromatographic elution profile is shown in Fig. 4. Substitution of sodium acetate with sodium chloride in buffer B resulted in a slight change in the elution profile of cytochrome P-450 and a decrease in resolution. Although sodium chloride displayed a better resolving property in the earlier portion of the chromatogram, poor results were obtained in the regions of peaks 4 and 5. As expected, lowering the sodium chloride

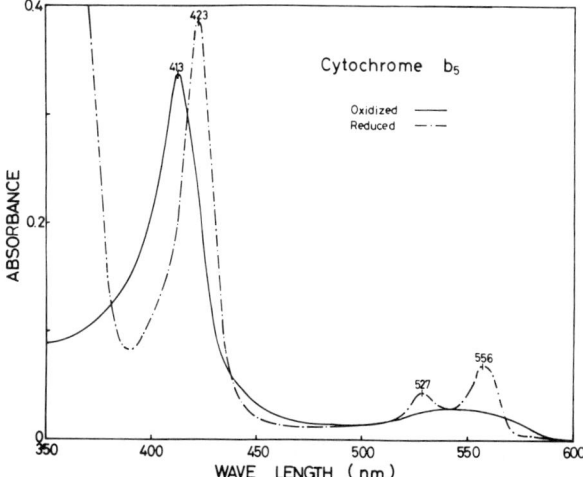

Figure 2. Absorption spectra of oxidized and reduced phenobarbital treated rat liver cytochrome b_5. Peak 8 shown in Fig. 1 is used as a sample. Buffer A of which composition is described in Fig. 1 is used as a reference buffer.

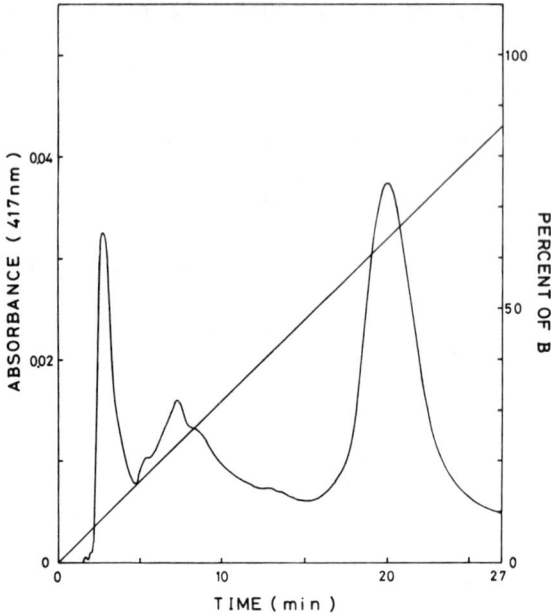

Figure 3. Effect of phosphate buffer. Sample is phenobarbital-induced rat liver microsome. Buffer A; 0.02 M potassium phosphate buffer, pH 7.8, containing 0.2% Emulgen 911 and 20% glycerol. Buffer B; 1.0 M sodium acetate is added to buffer A. pH 7.8. The other conditions are the same as in Fig. 1.

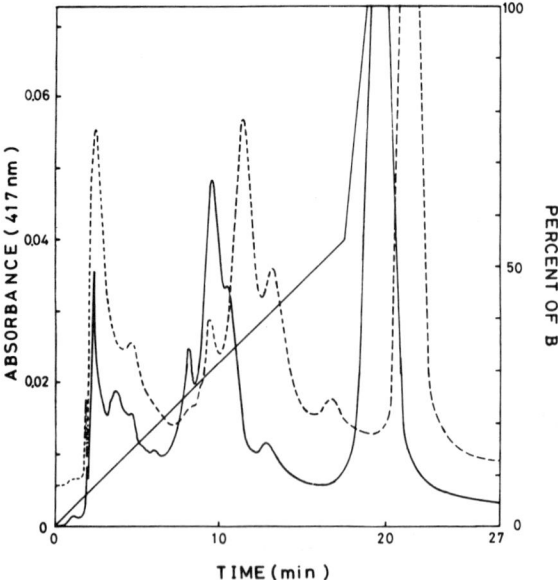

Figure 4. Effect of sodium chloride. Sample is phenobarbital-induced rat liver microsome. Buffer A; 0.02 M Tris-chloride, pH 7.2, containing 0.2% Emulgen 911 and 20% glycerol. Buffer B; 0.5 M (dotted line) or 0.8 M (straight line) sodium chloride is added to buffer A, pH 7.4. The other conditions are the same as in Fig. 1.

concentration to 0.5 M resulted in an increase in the retention time. Although the resolution of peaks 4 and 5 was better at the lower sodium chloride concentration, it was less satisfactory than that observed with sodium acetate.

Buffers containing 20% glycerol have been used to stabilize the activity of cytochrome P-450 (Ichikawa and Yamano, 1967). The addition of 20% glycerol to the mobile phase resulted in solutions of high viscosity. The use of high viscosity chromatography buffers lead to a decrease in resolution in gel-permeation chromatography and high column pressure. The effect of the concentration of glycerol on the HPLC profile of cytochrome P-450 was studied and the HPLC profile is shown in Fig. 5. Here the concentration of glycerol was 10%. The elution profile was identical with that observed with 20% glycerol buffer. Pump pressure of 1750–2200 psi with 20% glycerol in the mobile phase buffer was higher than the 1400 psi observed with 10% glycerol. Emulgen 911 is a non-ionic detergent commonly used in the solubilization and chromatographic separation of cytochrome P-450. Figure 6 shows the effects of various concentrations of Emulgen 911 dissolved in buffers on the chromatographic properties of cytochrome P-450. At the lower concentration of Emulgen 911 (0.05%, 0.1%), peaks 4 and 5 were not resolved and their retention times were longer than those observed with higher concentrations of Emulgen 911 (0.2%, 0.3%). No significant differences were observed between buffers containing 0.2% and 0.3% Emulgen 911. Since the enzymatic activity of cytochrome P-450 is inhibited by high concentrations of

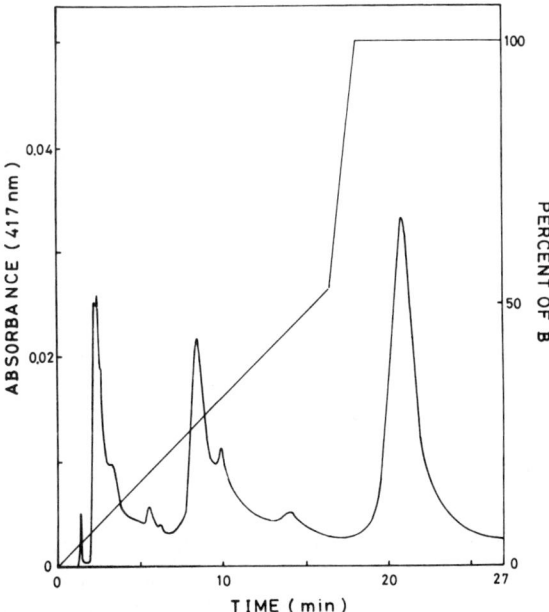

Figure 5. Effect of concentration of glycerol. Sample is phenobarbital-induced rat liver microsome. Buffer A; 0.02 M Tris-acetate, pH 7.3, containing 0.2% Emulgen 911 and 10% glycerol. Buffer B; 0.8 M sodium acetate is added to buffer A, pH 7.5. The other conditions are the same as in Fig. 1.

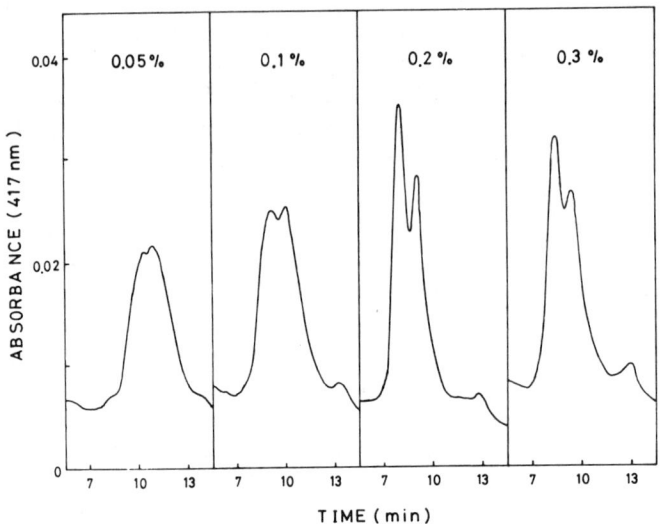

Figure 6. Effect of concentration of Emulgen 911. Sample is phenobarbital-induced microsome. Buffer A; 0.02 M Tris-acetate, pH 7.4, containing 0.05, 0.1, 0.2, and 0.3% Emulgen 911 and 20% glycerol. Buffer B; 0.8 M sodium acetate is added to buffer A, pH 7.4. The other conditions are the same as in Fig. 1.

Emulgen 911, 0.2% rather than 0.3% of the Emulgen was used in the majority of the reported studies.

Sodium cholate is another component widely used in the solubilization and isolation of cytochrome P-450. When sodium cholate was added to buffers A and B (0.5%, w/v), cytochrome P-450 and b_5 were not retained on the column. The ionic character of sodium cholate may be responsible for the decrease in ion-exchange capacity of the resin and would explain the poor retention of cytochrome P-450 on the column.

The porous character of the chromatographic support material increases the surface area and the ion-exchange capacity increases remarkably. Since the majority of the charges reside within the pores, penetration of the protein into the support is an important factor in the separation. We investigated the effects of pore size on the separation of cytochrome P-450 induced by PB pretreatment. The chromatographic profiles of cytochrome P-450 were examined using chromatographic supports with different pore sizes (100, 300, and 1000 Å).

The HPLC profiles of cytochrome P-450 using support materials with a pore size of 100 and 1000 Å are shown in Fig. 7. The HPLC profiles were fundamentally similar to the profile observed with 300 Å pore material (Fig. 1) but differed dramatically in resolution. With columns packed with 100 Å pore size support material, resolution of peak 1 was better than those observed in the later portions of the chromatogram. With the column packed with 1000 Å pore material, resolution of the peaks in the area of peaks 4—8 was better than in the early peaks in the chromatogram. However, neither support gave chromatograms of the quality

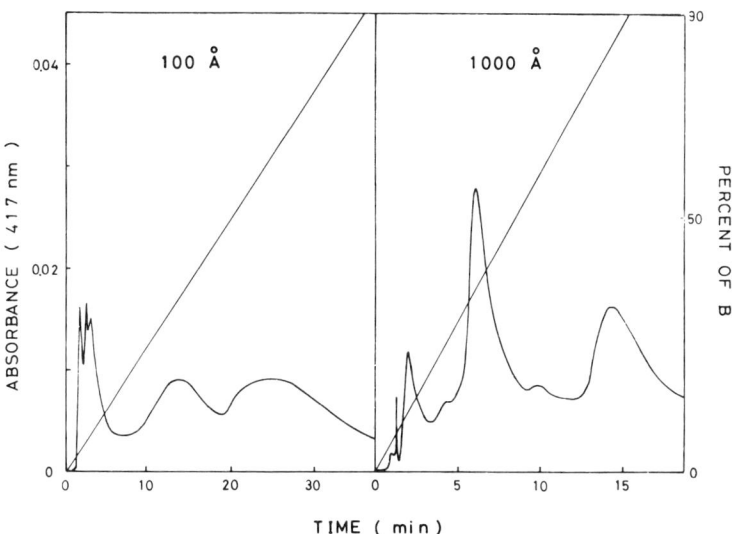

Figure 7. Effect of pore size of support material. Sample is phenobarbital-induced microsome. Analytical column, 4.6 × 120 mm i.d. stainless steel; pore size of support material, 100 and 1000 Å; particle size, 10 μm; temperature, 25°C; sample volume, 50 μl; detector, 417 nm. Left-hand (100 Å) buffer A; 0.02 M Tris-acetate, pH 7.9 containing 0.4% Emulgen 911 and 20% glycerol. Buffer B; 0.5 M sodium acetate is added to buffer A, pH 7.8. Flow rate, 0.5 ml/min; gradient rate, 100%/40 min; initial pressure, 530 psi. Right-hand (1000 Å) buffer A; 0.02 M Tris-acetate, pH 7.9, containing 0.2% Emulgen 911 and 20% glycerol. Buffer B; 0.8 M sodium acetate is added to buffer A, pH 7.3. Flow rate, 1.0 ml/min; gradient rate, 100%/15 min; initial pressure, 530 psi.

observed with the 300 Å support. The optimum pore size of the support material for resolving cytochrome P-450 is close to 300 Å.

HPLC profile of nontreated and PB, 3-MC, and PCB pretreated cytochrome P-450

Chromatograms of solubilized rat liver microsomes prepared from nontreated PB, 3-MC, and PCB pretreated rats are shown in Figs 1, 8–10. The chromatographic conditions are the same as described in Fig. 1. Eight major peaks were observed in the chromatographic profile of microsomes prepared from nontreated rats. If the small peaks and shoulders were included, 12 heme-containing components could be observed. Peak 8 eluted at a high concentration of salt was defined as cytochrome b_5, from spectral studies, as mentioned at the explanation of profile of PB treated cytochrome P-450 (Figs 1 and 2). The height of cytochrome b_5 peaks did not vary with treatment of different kinds of inducers (nontreated, PB, 3-MC, and PCB), indicating that cytochrome b_5 was hardly induced by these agents. The height of peaks 1–7 was small compared to that observed for peak 8 assigned to cytochrome b_5. This result coincides with findings when the specific content of nontreated cytochrome P-450 was low. In the profile of PB-induced solubilized microsomes, peaks 1 and 4 were increased with concomitant smaller increases in peak 5. The heights of other peaks except for 1, 4,

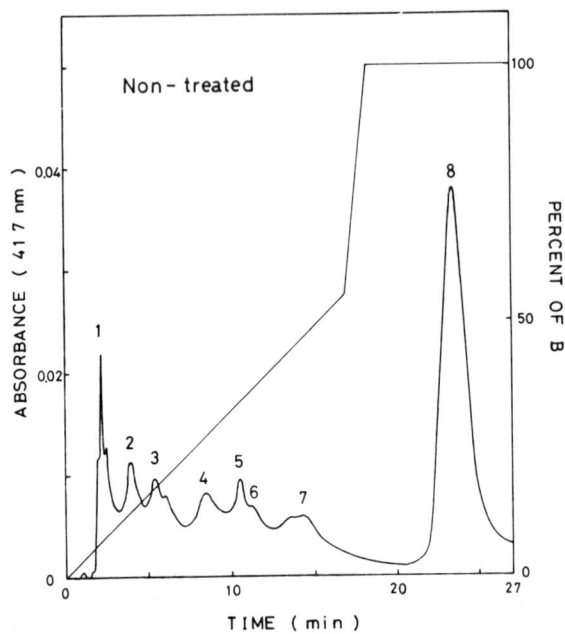

Figure 8. Profile of nontreated rat liver microsome. Conditions are the same as in Fig. 1.

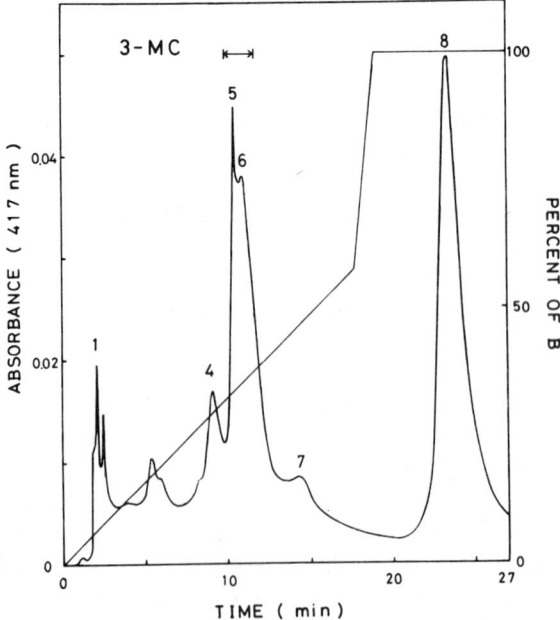

Figure 9. Profile of 3-methylcholanthrene-induced rat liver microsome. Conditions are the same as in Fig. 1. Fraction shown with arrow is collected for measuring of absorption spectra.

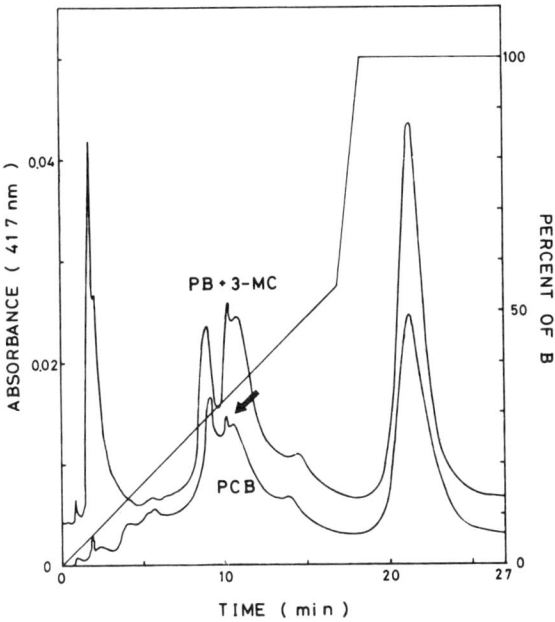

Figure 10. HPLC profile of PCB-induced cytochrome P-450. Microsomes were prepared by combining solubilized microsomes obtained from PB- and 3-MC-treated rats (PB + 3-MC) and microsomes prepared from PCB-treated rats (PCB). HPLC conditions are the same as in Fig. 1.

and 5 were similar to those observed in preparations of nontreated rats. Peak 1 was not stable, it decreased on standing at room temperature and remained as long as the samples are kept frozen. Treatment with 3-MC resulted in remarkable increases in peaks 5 and 6 with a concomitant slight increase in peak 4. Positions of peaks 2 and 3 were slightly different, however, peak height was similar to the nontreated. PCB treatment caused a clear increase in the two independent peaks shown in Fig. 10. One corresponded to peak 4 induced by PB treatment, determined from the data on the retention time. The other peaks corresponded to those obtained from the rat treated with 3-MC. The HPLC profile of PCB treated microsomes showed distinctly that the manner of induction of PCB was the mixed type of PB and 3-MC. This result confirmed data reported by other investigators (Alvares *et al.*, 1973; Ryan *et al.*, 1977; Parkinson *et al.*, 1980), all of whom used different methods.

Spectrum studies of cytochrome P-450 induced by treatment with PB, 3-MC, and PCB

In this experiment, 200 or 300 μl of solubilized microsomes containing about 5 nmol of cytochrome P-450 was sufficient for the measurement of maximum absorbance of the CO-reduced form. Following application of the samples on HPLC, cytochrome P-450, which was induced by different compounds (PB, 3-MC, and PCB), was collected. The absolute spectra of the oxidized, reduced, and CO-reduced forms were measured and the data are shown in Figs 11–13, respectively.

Maximum absorbance of CO-reduced form of peak 4 slectively induced with PB was observed at 450 nm. There are a small shoulder at 424 nm in the absolute spectrum not detected in CO-reduced difference spectrum. Peak 5 showed the same maximum absorbance as peak 4 at 450 nm. In the oxidized spectrum, maxima were seen at 418, 535, and 571 nm. Upon reduction of cytochrome P-450 by dithionite, the Soret band shifted to a lower wavelength (417 nm) and was accompanied by decrease in absorbance. A single band replaced the α and β bands seen in the oxidized spectrum. The shift of the Soret band toward the blue and the absence of a distinct peak or shoulder at 556 nm upon reduction provided

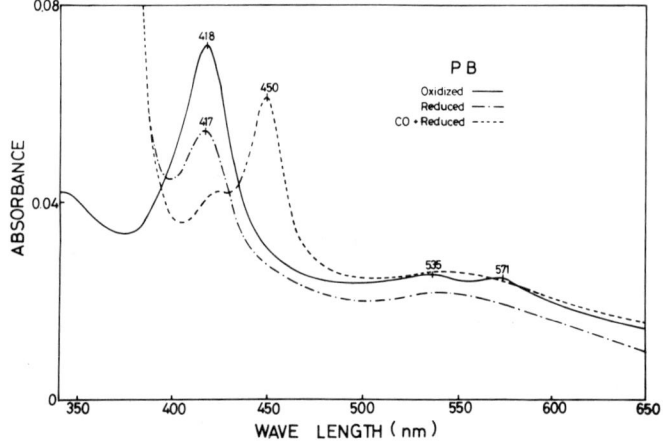

Figure 11. Absorption spectra of phenobarbital-induced cytochrome P-450. Sample solutions is peak 4 fraction shown in Fig. 1 without further concentration.

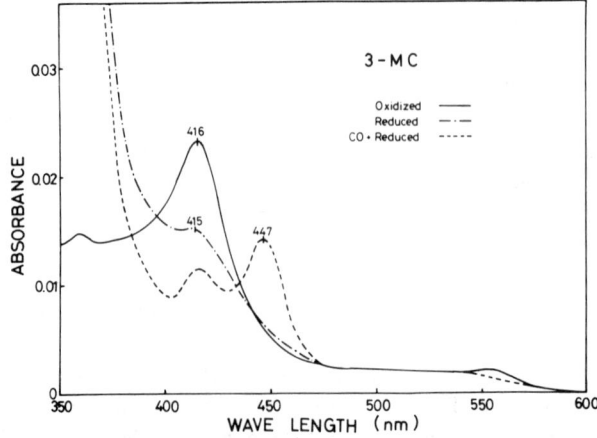

Figure 12. Absorption spectra of 3-methylcolanthlene-induced cytochrome P-450. Fraction shown with arrow in Fig. 9 is used as a sample solution without further concentration.

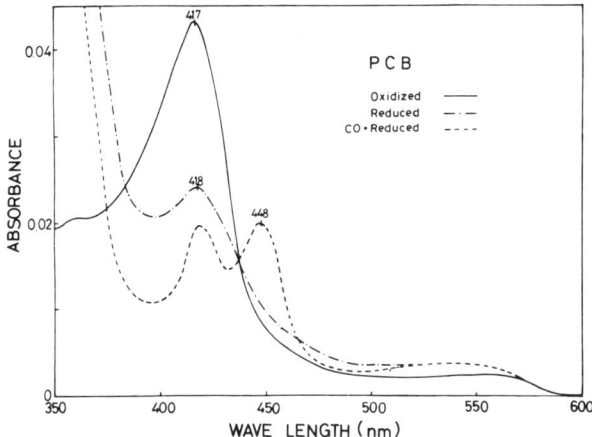

Figure 13. Absolute spectra of BCB-induced cytochrome P-450. Sample solution is the peak fraction selectively induced by PCB pretreatment shown with arrow in Fig. 10.

confirmatory evidence for the absence of cytochrome b_5. The absorption spectra of oxidized and reduced cytochrome b_5 from PB treated microsome are shown in Fig. 2. The oxidized form of the cytochrome P-450 showed an absorption maximum at 413 nm, and in the reduced form, the maxima appeared at 423, 527, and 556 nm, a typical spectrum of cytochrome b_5.

Absolute absorption spectra of cytochrome P-450 selectively induced by the treatment with 3-MC (peak 5 and 6) are shown in Fig. 12. The oxidized form showed a Soret absorption peak at 416 nm, indicating a low-spin form. In measurement of an absolute spectra, 0.2% Emulgen 911 is included in the cytochrome P-450 solution collected on HPLC, thus the Soret peak is at 416 nm not at 394 nm (Hashimoto-Yutsudo *et al.*, 1980) and α and β bands were not well resolved. In the reduced form, the Soret absorption peak was exhibited at 415 nm and only one peak was seen in the visible region. The CO-reduced maximum was seen at 447 nm. This value was in good agreement with the maximum of the B fraction from 3-MC treated rats reported by Ryan *et al.* (1975). A small peak appeared at 416 nm. The CO-reduced maximum of cytochrome P-450 induced by PCB was at 448 nm, an area situated between the value of PB (450 nm) and 3-MC (447 nm). This spectral data supported the idea that the induction manner of PCB was of the mixed type of PB and 3-MC. Parkinson *et al.* (1980) reported similar results using microsomal cytochrome P-450.

Multiple forms of rabbit hepatic and renal cytochrome P-450
HPLC profiles of microsomal cytochrome P-450 nontreated and treated with PB and 3-MC from rabbit liver and kidney are shown in Figs 14 and 15. Multiple forms of cytochrome P-450 were detected in the rabbit liver and kidney. In the nontreated rat and rabbit liver, an approximately equal number of forms of cytochrome P-450 were detected, but the retention time was slightly different between each species. The four major hemoprotein peaks were designated as

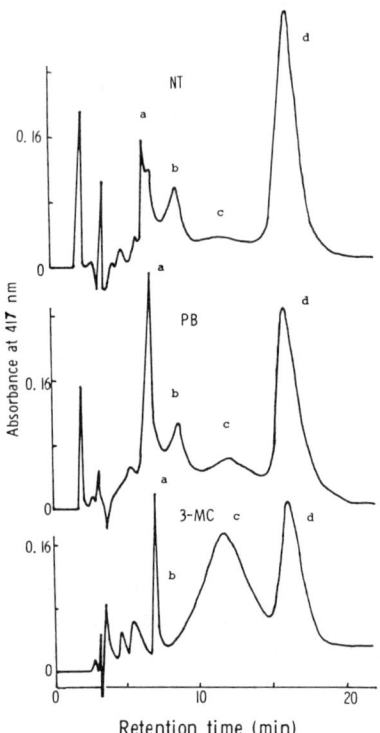

Figure 14. Profiles of rabbit liver microsomes. NT, PB, and 3-MC shows the HPLC profile of microsomes from nontreated, PB-treated, and 3-MC-treated rabbit liver, respectively. Mobile phase is shown in Fig. 1. The chromatogram was developed with a linear gradient from 0 to 0.8 M sodium acetate in 20 min at a flow rate of 1.0 ml/min.

peaks a, b, c, and d. Peak d was identified as cytochrome b_5. When rabbits were treated with PB, peak a was increased, while few changes were observed in the other peaks (Fig. 14). Treatment with 3-MC resulted in a great increase in peak c. Figure 15 shows the HPLC profiles of hemoproteins in solubilized kidney cortex microsomes from rabbits nontreated and treated with PB and 3-MC. Although the specific content of cytochrome P-450 from kidney is approximately ten times lower than that in the liver, several hemoprotein peaks were observed, even in the HPLC profile of kidney microsomes. Treatment with PB resulted in a nonsignificant changes in peak a, and peak b became a shoulder. Peak c was not detected in a significant amount in the case of nontreated and PB-treated microsomes, but following treatment with 3-MC, it was markedly increased. The CO-difference spectrum of peak c showed a maximum at 448 nm. Peak d was identified as cytochrome b_5.

Partial purification of different forms of cytochrome P-450 by HPLC
Generally speaking, HPLC shows extremely high resolution activity compared to conventional chromatographic methods, however, the sample loading capacity of

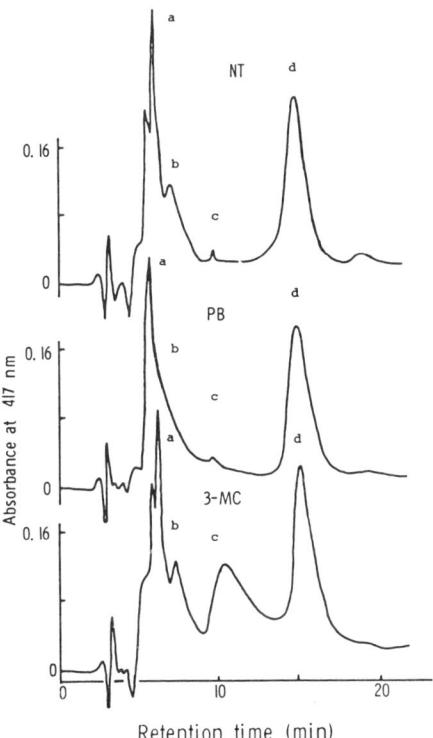

Figure 15. Profiles of rabbit kidney cortex microsome. NT, PB, and 3-MC shows the HPLC profile at microsomes from nontreated, PB-treated, and 3-MC-treated rabbit kidney cortex, respectively. HPLC conditions are the same in Fig. 14.

the former is lower. To purify large quantities of samples, we injected the same sample several times into HPLC column of regular size and manually collected the same fractions.

We attempted to purify two different forms of cytochrome P-450 induced by PB- and 3-MC-treatment and the results are the peak PB-4 (P-450PB) and peak MC-5, 6 (P-450MC) in Fig. 1 and Fig. 9, respectively. For the preparation of P-450PB, 400 μl of solubilized microsomes containing 11.8 mg of protein was applied on the HPLC. Following three injections, the collected P-450PB fractions were mixed. For preparation of P-450MC, 500 μl of solubilized microsomes induced by 3-MC-treatment containing 4.6 mg of protein was applied on the HPLC. Following four injections, the collected P-450MC fractions were mixed.

The viscosity of the solubilized microsomes induced by PB- and 3-MC-treatment was high as the concentration of protein was high, 29.5 and 9.1 mg/ml, respectively. Twenty percent of glycerol was included in the solubilized microsome. However, there were no untoward events except for a slight elevation of column pressure (less than 100 psi). Recovery of total cytochrome P-450 was over 70%.

The quantities of protein and cytochrome P-450 applied on HPLC, the specific

Table 1. Partial purification of P-450PB and P-450MC

Cytochrome P-450	Protein (mg)	Cytochrome P-450 (nmol)	Specific content (nmol/mg)	Absorption maximum (nm)	Molecular weight
Microsome (PB)	35.9	56.8	1.60	450	
P-450PB	1.13	6.92	6.12	450	51 000
Microsome (MC)	18.3	30.7	1.68	448	
P-450MC	0.80	7.31	9.13	447	54 000

HPLC of solubilized microsome induced by PB- and 3-MC-treatment was performed at a flow rate of 1.0 ml/min with a linear sodium acetate gradient (20%/30 min) made by controlling buffer A and B as described in the text. Molecular weight was calculated from the densitogram of SDS–PAGE.

content of cytochrome P-450, percent of recovery, absorption maximum, and molecular weight are given in Table 1. The laser densitogram of SDS–PAGE on P-450PB and P-450MC is shown in Fig. 16.

Twelve percent of the applied cytochrome P-450 was recovered in the P-450PB fraction. The specific content of P-450PB was 6.12. This value indicates that P-450PB from the solubilized microsomes was purified 3.8 times. On the densitogram, P-450PB gave a major protein-staining band with a molecular weight of

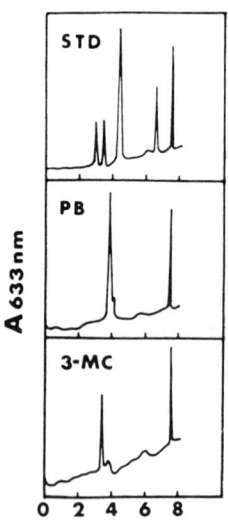

Migration Distance (cm)

Figure 16. Densitogram of SDS–PAGE. Electrophoresis was carried out in 9% acrylamide of 1 mm thick gel, described under Materials and Methods. Standard proteins (STD) are bovine serum albumin (M_r = 68 000), catalase (M_r = 58 000), ovalbumin (M_r = 45 000) and α-chymotripsinogen (M_r = 24 500). PB and 3-MC shows the SDS–PAGE densitogram of fractionated PB-4 (P-450PB) and MC-5 (P-450MC) shown in Figs 1 and 9, respectively.

51 000 and two faint bands with molecular weights of 49 000 and 47 000, respectively. The purity of the major band, calculated from the peak area of densitogram was 74%. The recovery of P-450MC from solubilized microsomes was 23.8%. The specific content of P-450MC was 9.13, which means that P-450MC was purified 5.4 times from the solubilized microsomes. On the densitogram, P-450MC gave major and faint bands corresponding in molecular weight to 54 000 and 51 000, respectively. The purity of this main band was calculated to be 67%. Therefore, P-450PB and P-450MC are different, as determined from data on the retention time of HPLC, maximum of CO-reduced form and molecular weight.

DISCUSSION

Ion-exchange liquid chromatography is commonly used to separate and purify proteins. However, this procedure has disadvantages. Considerable time is required to accomplish separation and hence the risk of obtaining denatured proteins and inactive enzymes is increased. The reproducibility of the chromatograms and resolution capabilities of this technique are poor and the procedure requires large sample quantities. A number of studies have been done on the separation of proteins by ion-exchange HPLC chromatography (Chang *et al.*, 1976; Regnier and Gooding, 1980; Regnier, 1982; Vacik and Toren, 1982). The advantages of the HPLC technique over conventional chromatography methods are (1) rapid manipulation, (2) high resolution, (3) high reproducibility, and (4) small sample quantity. Generally, the HPLC support materials used for the separation of protein are coated with an organic phase bound covalently to silica. This bonded phase deactivates the polar and charge groups on the surface of the silica. The ion-exchange properties are the result of the amino group covalently bonded to the support column. Separation of proteins by these support materials may depend upon three modes of chromatography. First, separation of proteins may be accomplished by ion-exchange. These support materials are made up of small porous particles which possess large surface areas and a high ion-exchange capacity. Second, the uniform pore size of the support may contribute to gel-permeation chromatography properties for separating proteins on the basis of molecular weights. Third, the organic surface coating material may contribute to the hydrophobic affinity properties of this support material. This mode of separation may be extremely useful for the separation of membrane-bound proteins which are hydrophobic in character. Due to similarities in the molecular weight of cytochrome P-450s, gel-permeation characteristics would contribute little to the separation of different forms of cytochrome P-450. However, ion-exchange and hydrophobic affinity properties of this support material may play a major role in the separation of different forms of cytochrome P-450.

Optimum conditions of HPLC

The feasibility of separating membrane-bound proteins by HPLC was investigated. Additionally, hepatic microsomes are restricted to two classes of hemoproteins, cytochrome P-450 and cytochrome b_5. The selection of the proper chromatographic conditions for separating a mixture of components is often a trial and error procedure. Many parameters were studied before adequate separation of solubilized rat liver cytochrome P-450 was accomplished. For the separation of cytochrome P-450, the most important factors are salt composition of the gradient buffer, pore size, and Emulgen 911 concentration. The resolution of cytochrome P-450 was more satisfactory using a sodium acetate salt gradient than with a sodium chloride gradient. In addition, the less corrosive properties of sodium acetate minimized possible damage to the HPLC equipment.

Effect of pore size
Pore size played an important role in the separation of cytochrome P-450. We compared 100, 300, and 1000 Å pore size support on the HPLC profile of cytochrome P-450. The surface area and ion-exchange capacity of the support material is inversely proportional to the support size. According to the data published by SynCrom Inc. for SynChrom Pac GPC permeation support (Linden, IN, USA) for soluble proteins, 100 Å pore supports would give the best chromatographic separation for the 40 000–60 000 molecular weight range. The molecular weight of monomeric cytochrome P-450 is 40 000–60 000, yet we found that 3000 Å and not the 100 Å pore size support gave the best resolution. This was not surprising since Warner *et al.* (1978) and Ingelman-Sandberg and Gustafsson (1977) observed that the molecular weight of cytochrome P-450, in the presence of detergent determined by gel-permeation chromatography, was 150 000 and 98 000, respectively, They postulated that solubilized P-450 may exist as a timer or dimer. When the support material with 100 Å pore size is applied for separation of cytochrome P-450, the inner surface area of porous particles where great ion-exchanging capacities are present, may not be available to such high molecular weight protein. The resolution of cytochrome P-450 on the 1000 Å pore size support was inferior to that observed with the 300 Å pore size support. However, it had a better resolving capacity than the 100 Å pore size support. The 1000 Å pore size support has a smaller surface area and ion exchange capacity compared to the 300 Å support. However, the larger pore size allows the solubilized cytochrome P-450 to penetrate, thus making feasible ion-exchange.

Effect of detergent in mobile phase
Buffers containing the non-ionic detergent, Emulgen 911, have been utilized in solubilization and ion-exchange chromatography of cytochrome P-450 (Imai and Sato, 1974; Warner *et al.*, 1978). Solubilized hepatic cytochrome P-450 is stable at room temperature in the presence of Emulgen (Warner *et al.*, 1978). The concentration of Emulgen 911 in the mobile phase played a critical role in the resolution of cytochrome P-450. For adequate resolution of multiple forms of cytochrome P-450, at least 0.2% of Emulgen 911 was required, and in this study this concentration was chosen. Emulgen 911 at concentrations over 0.2% in the mobile phase may be necessary to maintain the solubilization of proteins applied to the column. In addition, a sufficient quantity of the detergents may be required to decrease the hydrophobic property of the support, thereby enhancing the role of ion-exchange chromatography in the separation of cytochrome P-450.

Sodium cholate is commonly used for conventional DEAE-cellulose chromatography for separating cytochrome P-450. However, when sodium cholate was added to the elution buffer on the HPLC, cytochrome P-450 was not retained in the column. Sodium cholate is a kind of anionic detergent and therefore will react with the ion-exchanging group bound to the resin. Addition of sodium cholate to the mobile phase reduced the resolution of multiple forms of cytochrome P-450 in HPLC but not in conventional DEAE-cellulose chromatography (Warner *et al.*, 1978). The differences of the effects of sodium cholate with two different chromatographic methods may cause by the ion-exchange capacity and the chemical properties of support material. The concentration of Emulgen and sodium cholate in solubilization of microsomes did not fundamentally affect the HPLC profile. When over 2% Emulgen 911 was used for solubilization, peak 1 in Fig. 1 became slightly broad and without any change in the retention time.

Application for induction studies of cytochrome P-450

The HPLC profiles of induced cytochrome P-450 did not differ individually, but the magnitude of induction was slightly different, depending on the animals. According to our HPLC profiles, induction of rat hepatic microsomal cytochrome P-450 could be classified into three types: PB type, 3-MC type, and their mixed type. A typical example for the mixed type is the induction mode seen with PCB treatment. These three types were also supported from the value of CO-reduced maxima of selectively induced cytochrome P-450 by treatment with PB, 3-MC, and PCB, that is 450, 447, and 448 nm, respectively. Ryan *et al.* (1977) compared the SDS-polyacrylamide gel profiles, cross-reactivity with antibodies and spectral maxima of CO-reduced form of cytochrome P-450 from PB, 3-MC, and PCB treated rats. They reported that the properties of PB-treated cytochrome P-450 differed from those of 3-MC-treated ones and the properties of PCB-treated cytochrome P-450 belonged to the mixture of PB- and 3-MC-treated ones.

As shown in Fig. 1, peak 4 is mainly increased in the PB type. On the 3-MC type, peaks 5 and 6 are increased and peaks 4, 5, and 6 are increased simultaneously on the mixed type. Peak 4 of the HPLC profile of PB-induced microsomes locates between peaks 4 and 5 in the HPLC profile of nontreated microsomes. Peak 5 of the HPLC profile of the PB-induced microsomes corresponds to peak 5 of the HPLC profile of 3-MC-treated microsomes, determined from the value of the retention time. These data indicate that the anionic strength of cytochrome P-450 induced by 3-MC treatment is more potent than that induced by PB treatment. The similar elution profile was obtained by DEAE-cellulose chromatography of rat microsomes treated with PB and 3-MC (Elshourbagy and Guzelian, 1980).

HPLC profile of peaks 4 and 5 of cytochrome P-450 induced by PB was similar to the elution profile on DEAE-cellulose chromatography (Waxman and Walsh, 1982). From the order of elution profile, 'P-450PB' (termed by Waxman and Walsh) may correspond to PB-4. The molecular weight and elution order of peaks 4 and 5 also coincided with the results of fractions II and III obtained from PB-treated microsomes, respectively (Bornheim and Franklin, 1982). Sprague-Dawley male rats of the same strain were used throughout. Multiplicity of cytochrome P-450 was also investigated in the rabbit liver and kidney microsomes, using this HPLC technique (Figs 14 and 15). The molecular forms of cytochrome P-450 differ with the species (Guengerich *et al.*, 1981) and between the organs (Guengerich and Mason, 1979). In our studies, the HPLC profile of rabbit hepatic cytochrome P-450 differed from that of rats. Induction mode by the treatment with PB and 3-MC was similar to that seen in rats. Little information is available on the multiplicity of P-450 in extra-hepatic microsomes, in which the content of P-450 is much less than that in liver microsomes. Nevertheless, HPLC profiles of rabbit kidney microsomes showed a distinct multiplicity and induction mode by PB and 3-MC treatments. Since extrahepatic cytochrome P-450 is unstable, this HPLC technique which enables separation of a sample in a short time is suitable for the separation of extra-hepatic cytochrome P-450.

Partially purification of two different forms of cytochrome P-450

DEAE-cellulose chromatography is applied routinely for the separation of the multiple forms of cytochrome P-450, however, this method can detect three or four forms of cytochrome P-450 at most. Our HPLC technique for the separation of hemoprotein is a suitable method for studying multiple forms of cytochrome P-450 because the resolution activity of HPLC is very high, compared with conventional methods. Since the solubilized microsomes can be injected as a sample, modification of cytochrome P-450 which may occur during long procedure of purification can be excluded and the heterogeneity of intact cytochrome P-450 can be studied. The induction mode and magnitude by the treatment with various inducers are readily observed on the HPLC profile. Since each peak of hemoprotein is collected either manually or by using a fraction collector, identification of P-450 of each peak fraction can be made by measuring the CO-difference spectrum. In addition, the selectively induced cytochrome P-450 could be easily purified. Partially purified forms of cytochrome P-450 differ, as determined from the results of spectrum and retention time. The purity of cytochrome P-450s measured by the densitometry of SDS—PAGE profile was satisfied by single HPLC. However, the specific contents were low compared to the results using densitometry. This dissociation may be due to the assay of protein in the presence of Emulgen and other compounds which interfere with the Lowry assay method, as noted by Kohli *et al.* (1981) and Waxman and Walsh (1982). Cytochrome P-450 partially purified using the HPLC technique retained benzopyrene hydroxlating activity (Kusunose *et al.*, 1981). Thus, the HPLC technique can be used for preparation of cytochrome P-450 because the surface area of the support material is extremely high and there is a high ion-exchange capacity. We showed that the HPLC profile of ovalbumin did not change fundamentally even when 20 mg of sample was loaded on an analytical column (4.6 × 250 mm) (Funae *et al.*, 1982). The preparative column of the reverse-phase type and gel-permeation type are now commercially available.

CONCLUSIONS

This new HPLC technique is a rapid and excellent method which make feasible studies on the mode of induction of cytochrome P-450. Differences in multiple forms of P-450 can hardly be demonstrated using conventional methods, yet this HPLC method is suitable for such analyses, as there is a high resolution and reproducibility. Another advantage is that multiple forms of cytochrome P-450 can be prepared without fear of denaturation of the enzymatic activity.

ACKNOWLEDGEMENTS

We thank Carl Chatfield of Anspec for supplying the ion-exchange column. We also thank M. Ohara for helping us prepare this article and F. Tamura and R. Seo, C. Tsuda, K. Sato, and N. Yoneya for expert technique assitance. The laser densitometer was kindly donated by Dr and Mrs D. Sasaki. This research was supported by National Institutes of Health Grant, PHS GM-22220.

REFERENCES

Alvares, A. P., Bicker, D. R., and Kappas, A. (1973). Polychlorinated biphenyls: a new type of inducer of cytochrome P-448 in the liver. *Proc. Natl. Acad. Sci. USA 70*, 1321–1325.

Bornheim, L. M. and Franklin, M. R. (1982). Metabolic-intermediate complex formation reveals major changes in rat hepatic cytochrome P-450 subpopulations in addition to those forms previously purified after phenobarbital, β-naphthoflavone, and isosafrole induction. *Mol. Pharmacol. 21*, 527–532.

Chang, S. H., Gooding, K. M., and Regnier, F. E. (1976). High-performance liquid chromatography of proteins. *J. Chromatogr. 125*, 103–114.

Conney, A. H. (1967). Pharmacological implications of microsomal enzyme induction. *Pharmac. Rev. 19*, 317–366.

Elshourbagy, N. A. and Guzelian, P. S. (1980). Separation, purification and characterization of a novel form of hepatic cytochrome P-450 from rats treated with pregnenolone-16α-carbonitrile. *J. Biol. Chem. 255*, 1279–1285.

Funae, Y. and Kotake, A. N. (1982). Separation of multiple forms of rat liver cytochrome P-450 by high performance liquid chromatography. In: *Microsomes, Drug Oxidations and Drug Toxicity*, R. Sato and R. Kato (Eds.). Japan Scientific Societies Press, Tokyo, pp. 93–94.

Funae, Y., Seo, R., and Morimoto, S. (1982). High-performance liquid chromatography of proteins. *Japan. J. Clin. Chem. 11*, 342–347.

Funae, Y., Akiyama, H., Imaoka, S., Takaoka, M., and Morimoto, S. (1983). Rapid separation and measurement of rat urinary kallikrein by high-performance liquid chromatography with a continuous flow enzyme detector. *J. Chromatogr. 264*, 249–257.

Gooding, K. M., Lu, K. C., and Regnier, F. E. (1979). High-performance liquid chromatography of hemoglobins. *J. Chromatogr. 164*, 506–509.

Guengerich, F. P. (1978). Separation and purification of multiple forms of microsomal cytochromes P-450 partial characterization of three apparently homogeneous cytochrome P-450 isolated from liver microsomes of phenobarbital and 3-methylcholanthrene-treated rats. *J. Biol. Chem. 253*, 7931–7939.

Guengerich, F. P. (1979). Isolation and purification of cytochrome P-450, and the existence of multiple forms. *Pharmac. Ther. 6*, 99–121.

Guengerich, F. P. and Mason, P. S. (1979). Immunological comparison of hepatic and extra-hepatic cytochromes P-450. *Mol. Pharmacol. 15*, 154–164.

Guengerich, F. P., Wang, P., Mason, P. S., and Mitchell, M. B. (1981). Immunological comparison of rat, rabbit and human microsomal cytochrome P-450. *Biochemistry 20*, 2370–2378.

Hashimoto-Yutsudo, C., Imai, Y., and Sato, R. (1980). Multiple forms of cytochrome P-450 purified from liver microsomes of phenobarbital- and 3-methyl-cholanthrene-pretreated rabbits. II. Spectral properties. *J. Biochem. (Tokyo) 88*, 505–516.

Haugen, D. A. Van der Hoeven, T. A., and Coon, M. J. (1975). Purified liver microsomal cytochrome P-450: separation and characterization of multiple forms. *J. Biol. Chem.* *250*, 3567–3570.

Haugen, D. A. and Coon, M. J. (1976). Properties of electrophoretically homogenous phenobarbital-inducible and β-naphthoflavone-inducible forms of liver microsomal cytochrome P-450. *J. Biol. Chem.* *251*, 7929–7939.

Ichikawa, Y. and Yamano, T. (1967). Reconversion of detergent and sulfhydryl reagent-produced P-420 to P-450 by polyols and glutathione. *Biochim. Biophys. Acta.* *131*, 490–497.

Imai, Y. and Sato, R. (1974). A gel-electrophoretically homogeneous preparation of cytochrome P-450 from phenobarbital-induced rabbit liver microsomes. *J. Biochem. (Tokyo)* *75*, 689–697.

Ingelman-Sandberg, M. and Gustafsson, J. A. (1977). Resolution of multiple forms of phenobarbital-induced liver microsomal cytochrome P-450 by electrofocusing on granulated gels. *FEBS Lett.* *74*, 103–106.

Kohli, K. K., Linko, P., and Goldstein, J. A. (1981). Multiple forms of solubilized and partially resolved cytochrome P-450 from rats induced by 2,3,5,2′,3′,5′- and 3,4,5,3′,4′,5′-hexachlorobiphenyls. *Biochem. Biophys. Res. Commun.* *100*, 483–490.

Kohli, K. K., Hernandeg, O., and Mckinney, J. D. (1982). Fractionation by high performance liquid chromatography of microsomal cytochrome P-450 induced by hexachlorobiphenyl isomers. *J. Liq. Chromatogr.* *5*, 367–377.

Kotake, A. N. and Funae, Y. (1980). High-performance liquid chromatography technique for resolving multiple forms of hepatic membrane-bound cytochrome P-450. *Proc. Natl. Acad. Sci. USA* *77*, 6473–6475.

Kusunose, E., Kaku, M., Nariyama, M., Kusunose, M., Ichikawa, K., Funae, Y., and Kotake, A. N. (1981). High-performance liquid chromatography of cytochrome P-450 from rabbit liver, kidney cortex and intestinal mucosa microsomes. *Biochem. Int.* *3*, 399–406.

Laemmli, U. K. (1970). Cleavage of structural proteins during the assembly of the head of bacteriophage T_4. *Nature* *227*, 680–685.

Lowry, O. H., Rosebrough, N. J., Farr, A. L. and Randall, R. J. (1951). Protein measurement with the Folin phenol reagent. *J. Biol. Chem.* *193*, 265–275.

Lu, A. Y. H., Somogyi, A., West, S., Kuntzman, R., and Conney, A. H. (1972). Pregnenolone-16α-carbonitrile: a new type of inducer of drug-metabolizing enzymes. *Arch. Biochem. Biophys.* *152*, 457–462.

Omura, T. and Sato, R. (1964). The carbon monoxide-binding pigment of liver microsomes. I. Evidence for its hemoprotein nature. *J. Biol. Chem.* *239*, 2370–2378.

Parkinson, A., Cockerline, R., and Safe, S. (1980). Induction of both 3-methylcholanthrene- and phenobarbitone-type microsomal enzyme activity by a single polychlorinated biphenyl isomer. *Biochem. Pharmacol.* *29*, 259–262.

Phillips, A. H. and Langdon, R. G. (1962). Hepatic triphosphopyridine nucleotide-cytochrome c reductase: Isolation characterization and kinetic studies. *J. Biol. Chem.* *237*, 2652–2660.

Regnier, F. E. (1982). High-performance ion-exchange chromatography of proteins: the current status. *Anal. Biochem.* *126*, 1–7.

Regnier, F. E. and Gooding, K. M. (1980). High-performance liquid chromatography of proteins. *Anal. Biochem.* *103*, 1–25.

Reik, L. M., Levin, W., Ryan, D. E., and Thomas, P. E. (1982). Immunochemical relatedness of rat hepatic microsomal cytochromes P-450c and P-450d. *J. Biol. Chem.* *257*, 3950–3957.

Ryan, D., Lu, A. Y. H., West, S., and Levin, W. (1975). Multiple forms of cytochrome P-450 in phenobarbital and 3-methylcholanthrene-treated rats: separation and spectral properties. *J. Biol. Chem.* *250*, 2157–2163.

Ryan, D. E., Thomas, P. E., and Levin, W. (1977). Properties of purified liver microsomal cytochrome P-450 from rats treated with the polychlorinated biphenyl-mixture Aroclor 1254. *Mol. Pharmacol.* *13*, 521–532.

Ryan, D. E., Thomas, P. E., Korzeniowski, D., and Levin, W. (1979). Separation of multiple forms of highly purified liver microsomal cytochrome P-450 from rats treated with Aroclor 1254. *J. Biol. Chem. 254*, 1365–1374.

Schlabach, T. D., Alpert, A. J., and Regnier, F. E. (1978). Rapid assessment of isoenzymes by high-performance liquid chromatography. *Clin. Chem. 24*, 1351–1360.

Sladek, N. E. and Mannering, G. J. (1966). Evidence for a new P-450 hemoprotein in hepatic microsomes from methylcholanthrene treated rats. *Biochem. Biophys. Res. Commun. 24*, 668–674.

Thomas, P. E., Lu, A. Y. H., Ryan, D., West, S. B., Kawalek, J., and Levin, W. (1976). Multiple forms of rat liver cytochrome P-450: immunochemical evidence with antibody against cytochrome P-448. *J. Biol. Chem. 251*, 1385–1391.

Vacik, D. N. and Toren, E. C. (1982). Separation and measurement of isoenzymes and other proteins by high-performance liquid chromatography. *J. Chromatogr. 228*, 1–31.

Warner, M., Lamarca, M. V., and Neims, A. H. (1978). Chromatographic and electrophoretic heterogeneity of the cytochromes P-450 solubilized from untreated rat liver. *Drug Metab. Dispos. 6*, 353–362.

Waxman, D. J. and Walsh, C. (1982). Phenobarbital-induced rat liver cytochrome P-450. *J. Biol. Chem. 257*, 10446–10457.

Welton, A. F., O'Neal, F. O., Chaney, L. C., and Aust. S. D. (1975). Multiplicity of cytochrome P-450 hemoproteins in rat liver microsomes: preparation and specificity of an antibody to the hemoprotein induced by phenobarbital. *J. Biol. Chem. 250*, 5631–5639.

West, S. B., Huang, M. T., Miwa, G. T., and Lu, A. Y. H. (1979). A simple and rapid procedure for the purification of phenobarbital-inducible cytochrome P-450 from rat liver microsomes. *Arch. Biochem. Biophys. 193*, 42–50.

Progress in HPLC, Vol. 1, pp. 83—94
Parvez *et al.* (Eds)
© 1985 VNU Science Press

Gel permeation studies of glycoprotein by high-performance liquid chromatography

YASUYUKI SHIMOHIGASHI* and HAO-CHIA CHEN†
Endocrinology and Reproduction Research Branch, National Institute of Child Health and Human Development, National Institutes of Health, Bethesda, MD 20205, USA

INTRODUCTION

The gel permeation method has contributed greatly to the isolation, purification and characterization of biological macromolecules. Recently, the development of high-performance liquid chromatography (HPLC) and its application to gel permeation augmented their importance in separation sciences. The two most commonly used columns for gel permeation HPLC are the TSK G-SW and Synchropak GPC types. Their application to proteins (Okazaki *et al.*, 1980; Kato *et al.*, 1980a, b) and other biological macromolecules (Barth, 1980; Saito and Hayano, 1979; Stone and Krasowski, 1981) has been well documented. It is generally recognized that these columns provide several distinct advantages over the conventional soft-gel permeation columns such as Sephadex and Bio-Gel; namely, their significant reduction in both analysis time, sample size, band spreading, and increase of reproducibility and recovery values. For the isolation and purification of native or modified glycoproteins, gel permeation HPLC has been shown to be one of the most efficient procedures (Shimohigashi and Chen, 1982). Although gel chromatography (Ackers, 1975) and gel electrophoresis on sodium dodecyl sulfate-polyacrylamide (Weber and Osborn, 1975) have been widely used for the estimation of molecular weights of globular proteins because they are simple and accurate, these procedures when applied to glycoproteins often give a high estimate of molecular weight. This anomaly is attributed to the hydrophilic nature of carbohydrate moieties and the effect of its hydration on the molecular size (Andrews, 1965; Schubert, 1965; Bretscher, 1971). However, systematic studies on the direct relationship between carbohydrate content and chemical structures are lacking. With the better understanding of carbohydrate structures of glycoproteins and the advent of enzymatic desialylation and chemical deglycosylation methods, we

---

* Present address: Laboratory of Biochemistry, Faculty of Science, Kyushu University, Fukuoka, Japan.
† To whom correspondence should be addressed.

studied the elution behavior of human cholrionic gonadotropin (hCG), its subunits and their derivatives modified on carbohydrate structures on the gel permeation HPLC.

HCG is a glycoprotein hormone originated from placenta, which contains approximately 30% carbohydrate and is composed of two dissimilar α- and β-subunits (Birken and Canfield, 1980). In gel chromatography (Birken and Canfield, 1980) or gel electrophoresis (Chen, unpublished observations), hCG has shown apparent molecular weight 2–3 times larger than the actual values. Recently, the chemical structures of carbohydrates in hCG have been elucidated (Kessler *et al.*, 1979a, b; Endo *et al.*, 1979); namely, two different types of N–Asn-linked groups (n = 4) and O–Ser-linked (n = 4) as shown in Fig. 1. Progressive removal of carbohydrate residues by a series of exoglycosidases (Moyle *et al.*, 1975) and by chemical deglycosylation with anhydrous hydrofluoric acid-anisole (Sairam and Schiller, 1979; Chen *et al.*, 1982) has recently described. Employing hCG and its subunits as models, we have investigated the role of carbohydrate moieties in the apparent chromatographic anomaly of glycoproteins using gel permeation HPLC. The elution behavior of hCG and its subunits before and after deglycosylation was examined and the contribution of carbohydrate moieties to the overall size of these protein molecules was evaluated.

EXPERIMENTAL PROCEDURES

Proteins and reagents
Purified hCG and its α- and β-subunits (lot CR119) were obtained from the Center of Population Research, National Institute of Child Health and Human Development, National Institutes of Health. The following materials were used, all of which were HPLC grade: bovine thyroglobulin and soybean trypsin inhibitor (Sigma,

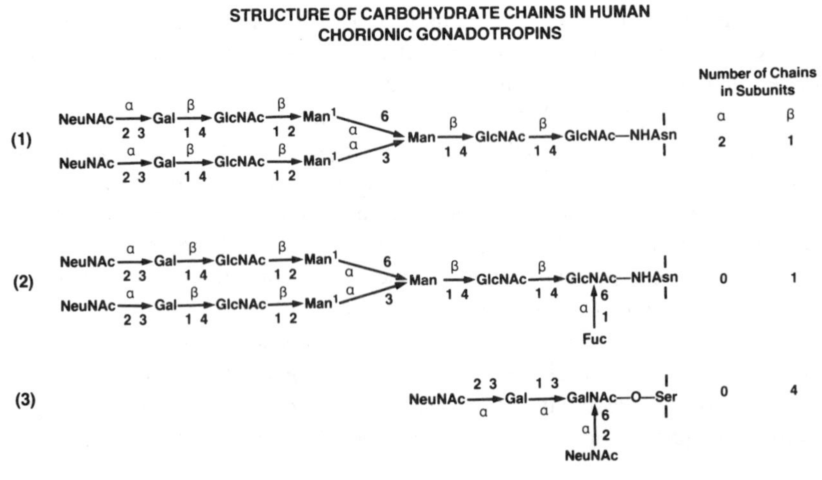

Figure 1. Structures of carbohydrate moieties in hCG. The abbreviations used are: NeuNAc, N-acetylneuramic acid; Gal, D-galactose; GlcNAc, N-acetyl-D-glucosamine; Man, D-mannose; Gal NAc, N-acetyl-D-galactosamine; Fuc, L-fucose; NH-Asn, asparaginyl.

St Louis, Mo.), bovine serum albumin (Miles, Elkhart, Ind.), ovalbumin (Pharmacia, Piscataway, N.J.) and phenylalanine (Pierce, Rockford, Ill.). Water, ammonium acetate (Fisher, Fairlawn, N.J.), and acetic acid (J. T. Baker, Phillipsburg, N.J.). The 0.1 M ammonium acetate–acetic acid buffer was filtered through a 0.45-μm filter (Millipore, Bedford, Mass.) before use.

Apparatus
Gel permeation HPLC was performed on a Hewlet-Packard 1084B liquid chromatograph (Hewlett-Packard, Avondale, Pa.), equipped with a processor-controlled sampling and u.v. monitoring systems. The amino acid analyzer Model 121MB and integrator Model 126 Data System were from Beckman Instument, Palo Alto, Calif.

Deglycosylation of hCG subunits
The removal of terminal sialic acids in the carbohydrate chains of hCG and its subunits (50 mg) was carried out by incubation (2 h, 37°C) with neuraminidase (EC 3.21.18, *Clostridium perfringens*, 92 μg of Type X; Sigma) in 0.5 M potassium acetate (pH 6.0), as previously described by Shimohigashi and Chen (1982). The pure, desialylated subunits (20 mg of each) were further treated with anhydrous liquid hydrofluoric acid (10 ml) for 1 h at 0°C in the presence of anisole as a scavenger. The complete removal of hydrofluoric acid and anisole was achieved by high-vacuum evaporation over potassium hydroxide pellets in a dessicator, followed by extraction of anhydrous solution of the product with diethylether.

Sephadex G-100 purification of deglycosylated hCG subunits
The purification of deglycosylated products (20–50 mg) was carried out by Sephadex G-100 gel chromatography on a series of four connected columns (3420 X 16 mm), previously equilibrated with 0.1 M ammonium bicarbonate (pH 7.9). In case of desialylation, the reaction mixture (*ca.* 2.5 ml) was directly applied to the column. The fractions containing subunits were pooled and twice lyophilized (*ca.* 40 mg). The hydrofluoric acid-treated asialo-subunits (*ca.* 20 mg) were dissolved in 0.1 M ammonium bicarbonate (2 ml) and chromatographed on the same series of Sephadex G-100 columns. The first peak to emerge (Fig. 2) was pooled and twice lyophilized (*ca.* 10 mg).

Gel permeation HPLC analysis of deglycosylated hCG subunits
Three TSK–GSW columns (each 600 X 7.5 mm: Beckman, Palo Alto, Calif.) were used in these studies. The samples of native hCG, its subunits and their deglycosylated derivatives (100–150 μg) were dissolved in 0.1 M ammonium acetate (150 μl). The solution was briefly centrifuged for 5–10 min in a rotary concentrator (Savant, Hicksville, N.Y.) and the clear supernatant was applied to the columns with an automatic sampling system connected to the HPLC apparatus. The elution conditions were as follows: buffer, 0.1 M ammonium acetate–acetic acid pH 7.0; flow rate, 1.0 ml/min; sample size, 100 μl; temperature, 25°C; u.v. detection, 278 nm.

Purification of hCG recombinants by gel permeation HPLC
Deglycosylated hCG subunits were recombined with each of the native complementary subunits by incubation (4°C, 24 h) in 0.1 M ammonium bicarbonate
(pH 7.9). The same amount of each subunit was employed for this recombination
as previously reported by Shimohigashi and Chen (1982). The reaction mixture was
lyophilized in a high-vacuum rotatory concentrator (Savant) in order to achieve the
maximal recombination of subunits by gradual concentration. The residue was
redissolved in 0.1 M ammonium acetate (150 μl) and purified in the gel HPLC
system consisting of one TSK-G2000SW and two G3000SW columns (total length,
1800 mm) connected in series. The fractions showing receptor-binding activity were
collected and lyophilized to yield 3–5 mg of pure products.

Analytical procedures
The amino acid analysis was performed in an amino acid analyzer after hydrolysis
of the sample in constant-boiling hydrochloric acid (110°C, 24 h). Amino sugars
were also determined by an amino acid analyzer after hydrolysis in 3 M hydrochloric
acid (100°C, 3 h). Quantitative analysis of sialic acid was carried out by the procedure of Warren (1975). Neutral sugars were determined according to the procedure
of Boykins and Liu (1980).

RESULTS AND DISCUSSION

The hCGα- and β-subunits were desialylated by neuraminidase and then deglycosylated by anhydrous hydrofluoric acid in the presence of anisole. The purification
of these modified subunits was achieved by Sephadex G-1000 gel chromatography.
Figure 2 shows a representative chromatogram of hydrofluoric acid-treated asialo-
hCGβ subunit, β(A–HF). Only the first peak was found to contain protein, whereas
peaks near the total permeation volume (fraction 130) contained the anisole-
carbohydrate conjugates derived from the hydrofluoric acid-catalyzed Friedel-Craft

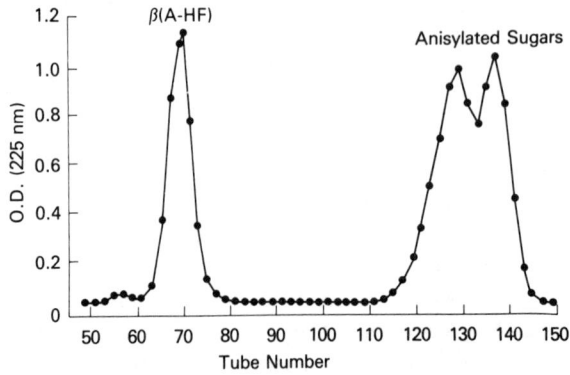

Figure 2. Elution pattern of HF-treated asiolo hCGβ on Sephadex G-100. Column, 3430 ×
16 mm; eluent, 0.1 M NH$_4$HCO$_3$ (pH 7.9); flow rate, 35 ml/h; fractions, 5.7 ml/tube; detection, 225 nm; temperature, 4°C; sample, 20 mg of HF-deglycosylated hCGβ.

reaction. Under the present conditions (0°C, 1 h), hydrofluoric acid treatment did not appear to alter polypeptide moeitity (Sairam and Schiller, 1979; Mort and Lamport, 1977). Amino acid analyses of all modified subunits after purification, indeed, revealed that their compositions were in close agreement with the expected stoichiometry of unmodified subunits (data not shown).

When the elution position of hCG subunits on a Sephadex column was compared with respect to the degree of deglycosylation, it was obvious that these modifications considerably affected the molecular size of glycoproteins (Chen *et al.* 1982). The partition coefficients for the α-subunits were 0.33 (asialo) and 0.53 (asialo-HF), and those for the β-subunits were 0.29 (asialo) and 0.38 (asialo-HF). The increased values of hydrofluoric acid-treated asialo subunits suggest that the molecular size of the glycoprotein was significantly reduced by the deglycosylation.

The results of sugar content analyses are shown in Table 1. The thiobarbituric quantitation method of Warren revealed the complete removal of sialic acid residues after the neuraminidase treatment. The complete desialylation corresponds to a reduction of 20–21% in carbohydrate content, and of 9% in molecular weight of both α- and β-subunits. Reductions caused by the hydrofluoric acid deglycosylation were more drastic: a 77–79% reduction of carbohydrate content and a 23–24% reduction of molecular weight of the two subunits (Table 2). No cleavage of N–Asn-linked N-acetyl-glucosamine or O–Ser-linked N-acetylgalactosamine was evident under the conditions used for the hydrofluoric acid treatment. As previously reported (Sairam and Schiller, 1979; Chen *et al.*, 1982), hydrofluoric acid appears to cleave predominantly the saccharide linkage at the mannose residues of the N–Asn-linked chain and at the galactose residue of the O–Ser-linked chain. Neither galactose nor fucose were detected in the two subunits after the treatment.

The elution behavior of these modified glycoproteins and standard globular proteins was studied by using TSK G-SW columns in an HPLC apparatus. Although the gel used in HPLC contains a large number of theoretical plates per unit volume, the currently available columns have one notable limitation: a small gel volume. In order to improve the resolution, an increase in the column length in addition to a reduction of flow rate can be the most advantageous method. Consequently, we studied the elution behavior of hCG and its α- and β-subunits in a series of column lengths at a flow rate of 1 ml/min as shown in Fig. 3. The use of a single column of TSK G2000SW or G3000SW (Fig. 3A) gave no separation between the native hCG and its β-subunit. Similarly, no satisfactory separation was indicated when two columns consisting of G2000SW and G3000SW were connected in series (Fig. 3B). However, when one G2000SW and two G3000SW columns were connected in series at a total length of 1800 mm, the separation of native hCG and its two subunits was accomplished (Fig. 3C). The excellent separation of the following reference proteins: bovine thyroglobulin (mol. wt 669 000), bovine serum albumin dimer (134 000) and its monomer (67 000), ovalbumin (43 000), soy bean trypsin inhibitor (21 000), and phenylalanine (169) were demonstrated using the same set of columns (Fig. 4). At a flow rate of 1 ml/min, the peak representing each component was symmetrical and they were almost completely resolved. The elution time in five replicate analyses were highly reproducible, with a standard deviation of 0.05 min, and a chromatogram can be completed in 75 min.

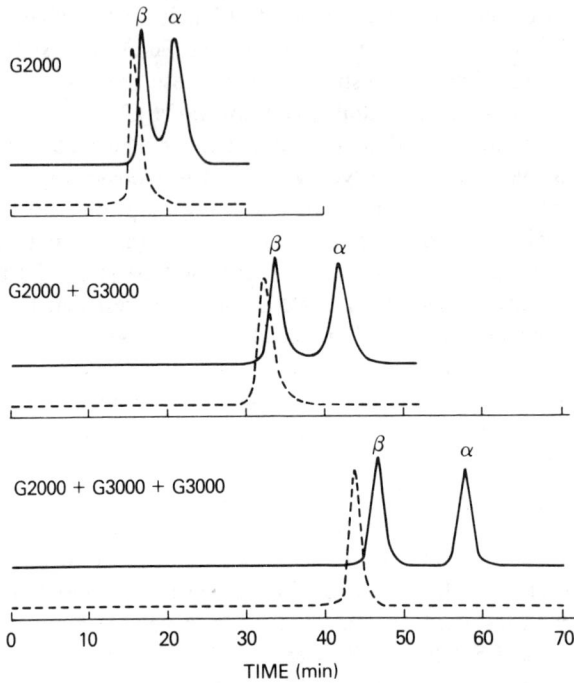

Figure 3. Gel HPLC elution profiles of hCG and its subunits in the different column length systems. ———, native hCG; and ———, mixture of hCGα- and β-subunits in 6 M guanidine HCl.

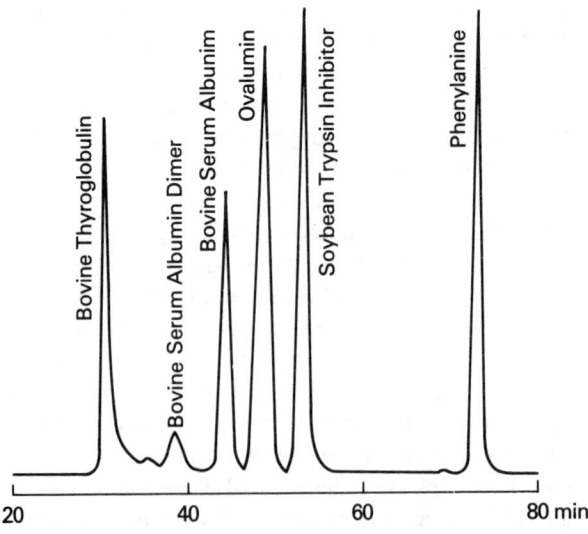

Figure 4. Standard gel HPLC elution pattern of globular proteins. Columns, TSK-G2000SW + G3000SW + G3000SW (in that order, 1800 × 7.5 mm); eluent, 0.1 M NH₄OAc–HOAc (pH 7.0); flow rate, 1.0 ml/min; sample, proteins (1.0 mg each) in 100 μl.

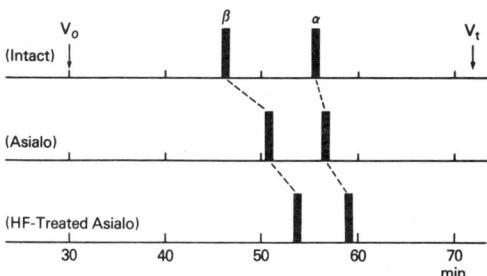

Figure 5. Elution positions of hCG subunits before and after deglycosylations. (Conditions as Fig. 4.)

When the native and deglycosylated hCG subunits were analyzed by gel HPLC, all subunits exhibited a single, symmetrical peak. Figure 5 shows the elution positions of the α- and β-subunits before and after deglycosylation. The two subunits with the same modification were well separated from each other, the β-subunit being eluted first because of its larger molecular weight. By progressive deglycosylations, the elution position of each subunit distinctly shifted to greater elution volumes: the differential partition coefficients, ΔK, caused by desialylation was 0.02 (α) and 0.09 (β), and ΔK caused by hydrofluoric acid deglycosylation was 0.07 (α) and 0.17 (β). The terms for partition coefficient (K) and ΔK were determined by use of the following equations

$$K = (V_e - V_0)/(V_t - V_0)$$

$$\Delta K = K_{(deglycosylated)} - K_{(native)}$$

where V_e is the solute elution volume. The void volume (V_0) and total volume (V_t) were determined as the elution volumes of bovine thyroglobulin and of amino acids or water, respectively. It should be noted that the shifts caused by each modification are more marked for the β- than for the α-subunit. The ΔK values for the β-subunit as compared to those of the α-subunit were larger by a factor of 0.07 and 0.10 after desialylation and hydrofluoric acid deglycosylation, respectively. Moreover, the large shift presumably become apparent after hydrofluoric acid deglycosylation because of a drastic reduction of carbohydrate content or molecular weight, as indicated above.

Since the progressive deglycosylation of hCG subunits reduces the molecular size accordingly, we have examined the effect of carbohydrate modifications of one subunit in the recombinant after association with its native complementary subunit. The separations by gel HPLC were excellent for the α- and β-subunits. However, the separation of hCG from the β-subunit was less satisfactory and worsened further when the hCG recombinants contained the deglycosylated subunits. In order to eliminate contamination of the free native or modified β-subunit in the purified recombinants the recombination reactions were carried out in the presence of excess α-subunits (Shimohigashi and Chen, 1982). This resulted in the formation of recombinants having an amino acid composition in close agreement with that of the intact hCG. As shown in Fig. 6, all recombinants were well separated

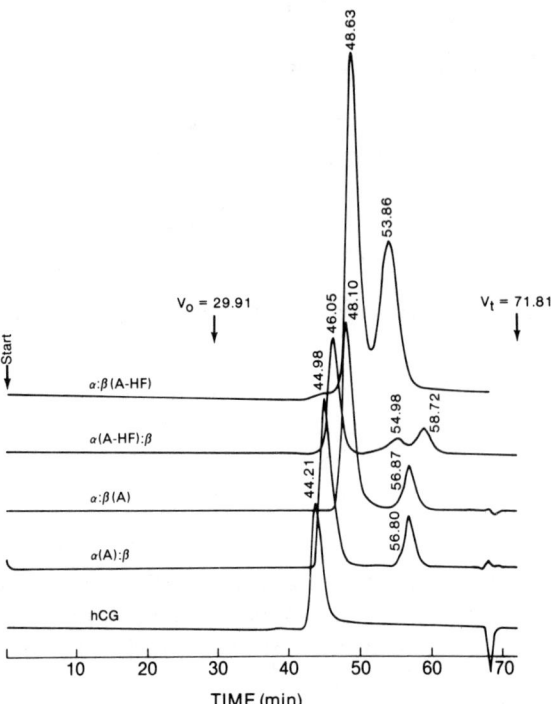

Figure 6. Gel-HPLC purification of hCG recombinants. Samples, products of recombination (3 mg of each subunit) in 150 μl. (Conditions as Fig. 4.)

from excess α-subunits. The greater shift in elution time was evident for the recombinants having hydrofluoric acid deglycosylated subunits. The downshift in elution time due to modifications of the β-subunit appear to be larger than those due to α-subunit modifications (Fig. 7). Overall, we have observed that the β-subunit exhibits the most pronounced change in elution behavior induced by the deglycosylations.

Like conventional gel chromatography, gel HPLC can also be applied to molecular weight determinations (Kato *et al.*, 1980; Gruber *et al.*, 1979; Ui, 1979).

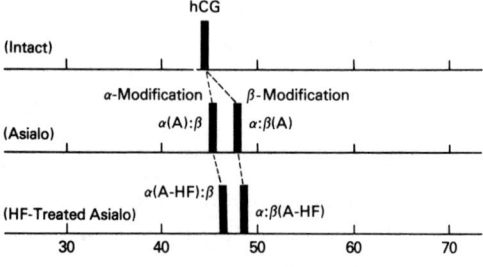

Figure 7. Elution positions of hCG recombinants with α- and β-subunit modifications. (Conditions as Fig. 4.)

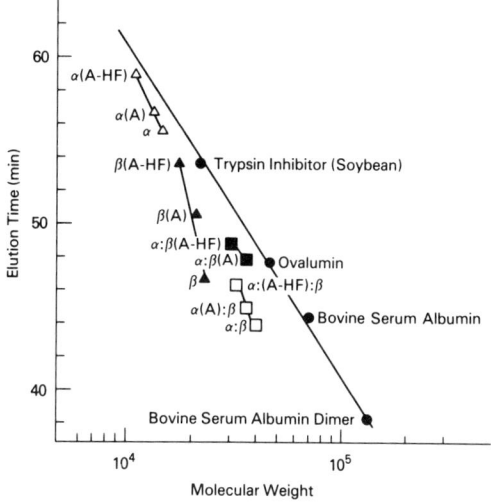

Figure 8. Correlation between molecular weight and elution time in gel HPLC. Molecular weights of hCG recombinants were calculated on the basis of those of the subunits shown in Table 2.

Figure 8 shows that an excellent straight-line relationship was obtained from the present gel HPLC system when the molecular weight on a logarithmic scale was plotted against the elution time of a series of standard globular proteins. Molecular weights (M_0) of the derivatives of hCG and its subunits were calculated on the basis of their carbohydrate and amino acid compositions (Tables 1 and 2). When the elution times of hCG proteins were plotted against their molecular weights, all subunits and hCG considerably deviated from the standard correlation (Fig. 9). Such deviations are most notable for the β-subunits. Apparent molecular weights (M) estimated by the standard calibration were larger than the calculated ones (M_0). The ratios of M/M_0 were 2.27 (native), 1.58 (asialo) and 1.30 (asialo-HF) for the β-subunit. For the α-subunit, the M/M_0 ratios, 1.26 (native), 1.21 (asialo), and 1.13 (asialo-HF), were lower than the corresponding β-subunits. These results clearly indicate that the differences between the calculated and apparent molecular weights were attributable to the change in carbohydrate content. The progressive removal of carbohydrate residues significantly reduced the deviation of molecular weight from the standard calibration curve, as shown by the ratio of M/M_0. When the differential deviation was calculated by the equation

$$\Delta(M/M_0) = M/M_0\,(\text{native}) - M/M_0\,(\text{deglycosylated})$$

the values indicated the extent to which the reductions in carbohydrate content cause the values to approach the standard calibration. The calculated values of differential deviation were 0.05 (α) and 0.69 (β) after desialylation, and 0.13 (α) and 0.97 (β) after hydrofluoric acid deglycosylation. It is noteworthy that such reversals of molecular weight by deglycosylation are quite dramatic for the β-subunit.

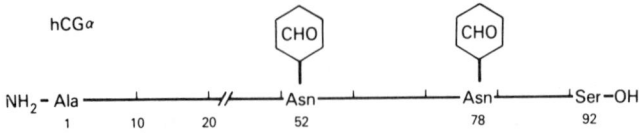

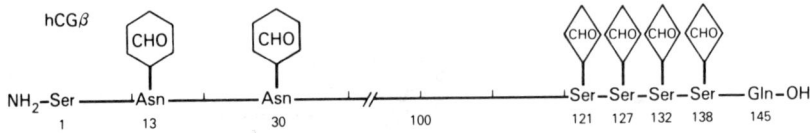

Figure 9. Location of carbohydrate moieties (CHO) in the α- and β-subunits of hCG.

Table 1. Monosaccharide composition of deglycosylated hGC subunits

	hCGα		hCGβ	
Sugar	asialo	asialo-HF	asialo	asialo-HF
NeuNAc	0	0	0 0	
Gal	3.0	0	6.3	0
Man	5.6	1.6	5.2	1.5
Fuc	—	—	1.2	0
GlcNAc	7.3	3.0	7.1	2.6
GalNAc	—	—	2.8	2.1
Total	15.9 (79%)	4.6 (23%)	22.6 (78%)	6.2 (21%)

Values are expressed as moles of monosaccharide per mole of protein and those in parentheses represent percentage of carbohydrate remaining after deglycosylations. Amounts of NeuNAc in native hCGα and hCGβ were 4.2 and 6.3 mol/mol protein, respectively.

Table 2. Calculated molecular weights of native and deglycosylated hCG subunits

	hCGα	hCGβ
Native	14 500 (30%)	22 200 (30%)
Asialo	13 200 (23%)	20 300 (24%)
Asialo-HF	11 000 (7.4%)	17 100 (9.3%)

The percentage values in parentheses represent the carbohydrate content. All values were calculated on the basis of compositions of polypeptide and carbohydrate from Table 1.

Similarly, native hCG had a ratio of 1.83, which was reduced to 1.6–1.7 by α-subunit modifications, and to 1.2–1.3 by β-subunit modifications (Fig. 9). These results indicate that the apparent molecular weight of recombinants containing the native β-subunit are uniformly higher than those for the α-subunit, and are influenced to a lesser extent by the α-subunit modifications. These data further suggest that the effective molecular size of hCG is predominatly determined by the β-subunit.

As to the molecular weight determination with soft-gels or gel electrophoresis, it is well known that glycoproteins give a significantly higher value if the estimation is based on the standard calibration curve derived from globular proteins (Ackers, 1975; Weber and Osborn, 1975; Andrews, 1965; Schubert, 1965; Bretscher, 1971; Birken and Canfield, 1980). For example, hCG, containing *ca.* 30% carbohydrate, was eluted at the position of globular proteins with an apparent molecular weight of 65 000–70 000, a considerable deviation from the value of 36 700 (Birken and Canfield, 1980). This anomalous elution behavior was also observed in the gel-HPLC system, where the molecular weight of hCG appeared to be 1.83 times larger than the true value obtained from the composition analyses. Similarly, the α- and β-subunit gave values 1.26 and 2.27 times, respectively, higher than the actual molecular weight.

In the present studies, the reduction of the carbohydrate content in the two subunits from 30% to 8% by two deglycosylation processes resulted in a decrease of the M/M_0 ratio from 1.26 and 2.27 to 1.13 and 1.30 for the α- and β-subunits, respectively. These data directly demonstrated the effect of carbohydrates on the anomalous elution behavior of glycoproteins. However, the question can be raised why the two subunits, both of which contain 30% of carbohydrate and had been deglycosylated to the same extent, exhibited such a difference in the reduction of the M/M_0 ratio after the deglycosylation. Previously, Andrews (1965) found no apparent correlation between the extent of deviation and the content of carbohydrate in glycoproteins. Moreover, ovalbumin showed no deviation from the reference globular proteins, as shown in Fig. 9, even though it contains 3.5% carbohydrate. Perhaps not only the carbohydrate content but also the location of carbohydrate chains in the spatial arrangement of the protein molecule determines the overall structural dimensions of glycoproteins. If this is the case, the high values obtained from hCGβ and its derivatives may be the result of the contribution by the unique carboxyl-terminal peptide, which contains four additional O–Ser-linked carbohydrate chains not found in the α-subunit as shown in Figs 1 and 9. It is likely that the carboxyl-terminal glycopeptide extends outward as an appendage (Ohashi *et al.*, 1980) and thus amplifies the overall molecular size. Our data also illustrate that the β-subunit is the dominant factor in determining the carbohydrate effect on the molecular dimensions of recombinants. It is intriguing that the carbohydrate moieties in the β-subunit also play a dominant role in maintaining the hormonal activities of the native and recombined molecule (Shimohigashi and Chen, 1982).

REFERENCES

Ackers, G. T. (1975). In: *The Proteins*, H. Neurath and R. L. Hill (Eds). 3rd edn, Vol. 1, Academic Press, New York (p. 1).
Andrews, P. (1965). *Biochem. J. 96*, 595.
Barth, H. G. (1980). *J. Liq. Chromatogr. 3*, 1481.
Birken, S. and Canfield, R. E. (1980). In: *Chorionic Gonadotropin*, S. J. Segal (Ed.). Proceedings

of Conference on Human Chorionic Gonadotropin, Bellagio, Italy, 14–16 November 1979. Plenum Press, New York (p. 65).

Boykins, R. A. and Liu, T. Y. (1980). *J. Biochem. Biophys. Methods 2*, 71.

Bretscher, M. S. (1971). *Nature, New Biol. 231*, 229.

Chen, H.-C., Shimohigashi, Y., Dufau, M. L., and Catt, K. J. (1982). *J. Biol. Chem. 257*, 1446.

Endo, Y., Yamashita, K., Tachibana, Y., Tojo, S., and Kobata, A. (1979). *J. Biochem. 85*, 669.

Gruber, K. A., Whitaker, J. M., and Morris, M. (1976). *Anal. Biochem. 97*, 176.

Kato, Y., Komiya, K., Sasaki, H., and Hashimoto, T. (1980a). *J. Chromatogr. 193*, 458.

Kato, Y., Komiya, K., Sawada, Y., Sasaki, H., and Hashimoto, T. (1980b). *J. Chromatogr. 190*, 305.

Kato, Y., Komiya, K., Sasaki, H., and Hashimot, T. (1980c). *J. Chromatogr. 190*, 297.

Kessler, M. J., Reddy, M. S., Shal, R. H., and Bahl, O. P. (1979a). *J. Biol. Chem. 254*, 7901.

Kessler, M. J., Mise, T., Ghai, R. D., and Bahl, O. P. (1979b). *J. Biol. Chem. 254*, 7909.

Mort, A. J. and Lamport, D. T. A. (1977). *Anal. Biochem. 82*, 289.

Moyle, W. R., Bahl, O. P., and Marz, L. (1975). *J. Biol. Chem. 250*, 9163.

Ohashi, M. Matsuura, S., Chen, H.-C., and Hodgen, G. D. (1980). *Endocrinology 107*, 2034.

Okazaki, M., Ohno, Y., and Hara, I. (1980). *J. Chromatogr. 221*, 257.

Sairam, M. R. and Schiller, P. W. (1979). *Arch. Biochem. Biophys. 197*, 294.

Saito, Y. and Hayano, S. (1979). *J. Chromatogr. 177*, 390.

Schubert, D. (1965). *J. Mol. Biol. 53*, 305.

Shimohigashi, Y. and Chen, H.-C. (1982). *FEBS Lett. 150*, 64.

Stone, R. G. and Krasowski, J. A. (1981). *Anal. Chem. 53*, 736.

Ui, N. (1979). *Anal. Biochem. 97*, 65.

Warren, L. (1975). *J. Biol. Chem. 234*, 1971.

Weber, K. and Osborn, M. (1975). In: *The Proteins*, H. Neurath and R. L. Hill (Eds). 3rd edn, Vol. 1, Academic Press, New York (p. 179).

Progress in HPLC, Vol. 1, pp. 95–103
Parvez *et al.* (Eds)
© 1985 VNU Science Press

Gel permeation of human serum lipoprotein by HPLC

ICHIRO HARA[1] and MITSUYO OKAZAKI[2]

[1] Scientific Instrument Division, Toyo Soda Mfg Co., Hayakawa, Ayase-shi, Kanagawa Prefecture 252, Japan and [2] Laboratory of Chemistry, Department of General Education, Tokyo Medical and Dental University Kohnodai, Ichikawa, Chiba Prefecture 272, Japan

INTRODUCTION

Human serum (plasma) lipoproteins were fractionated into four or five classes using the ultracentrifugation technique of de Lalla and Gofman (1954). The ultra-centrifugal fractionation of serum lipoprotein is known to be based on the difference of the specific gravities among serum lipoproteins; the name of each fraction is originated from the behavior in the ultracentrifugal field.

SERUM LIPOPROTEIN CLASSES AND THEIR PROPERTIES

The composition and the properties of serum lipoproteins is shown in Table 1. The particle sizes of serum lipoprotein fractions are quite different from each other resulting in the application of gel permeation chromatography using agarose gel by Sata *et al.* (1970) and hydroxyapatite by Kostner *et al.* (1977). However, gel permeation column chromatography takes a long time to complete.

High-performance aqueous gel permeation liquid chromatography (HPAGPLC) has been widely used in biochemistry because of its short time performance and higher reproducibility.

HIGH-PERFORMANCE AQUEOUS GEL PERMEATION LIQUID CHROMATOGRAPHY

Various kinds of columns for HPAGPLC are commercially supplied as shown in Table 2. It is clear that almost all particle sizes of the substances of high molecular weights using two series of TSK GEL PW and that of SW (Toyo Soda, Tokyo) are covered.

The reference standards of human serum lipoprotein are prepared by the sequential ultracentrifugal procedure of Havel *et al.* (1955). Each reference standard is assayed for its homogeneity by immunoelectrophoresis.

The reference standards are applied to the column system of TSK GEL G5000PW

Table 1. Properties and chemical compositions of human serum lipoproteins

	Chylomicron	VLDL	LDL	HDL$_2$	HDL$_3$
Density	0.95	0.95–1.00	1.006–1.063	1.063–1125	1125–1210
Molecular weight	0.4×10^9	$5–10 \times 10^6$	2.7×10^6	3.7×10^5	1.8×10^5
Particle size (Å)	750	250–750	200–300	60–140	40–100
Electrophoretical mobility	origin	prebeta	beta	alpha	alpha
Composition					
protein (%) lipid mg/100 mg lipoprotein lipid	0.5–2.5	10	25	45	48
TG	87.7	55.7	7.3	6.1	6.7
TC	3.0	16.8	59.2	42.5	38.4
PL	8.8	19.3	27.8	42.4	40.9
Apoprotein					
major	Apo B Apo C-I C-II C-III	Apo B Apo C-I, C-II, C-III Apo E	Apo B	Apo A-I, A-II	Apo A-I, A-II
minor	Apo A-I A-II	Apo A-I A-II Thin line pro.		Apo C-I, C-II, C-III Apo E	Apo C-I C-II, C-III Apo E

Table 2. Specifications of TSK GEL PW series and SW series

Type	Particle size (μm)	Molecular weight ranges		Theoretical plate number
		polyethylene glycol	dextran	
G2000PW	10	–5000		5000
G3000PW	13	–20 000	60 000	5000
G4000PW	13	500–30 000	1000–700 000	3000
G5000PW	17		50 000–2 000 000	3000
G6000PW	17		500 000–30 000 000	3000
G2000SW	10	5000–100 000	1000–30 000	>5000
G3000SW	10	10 000–500 000	2000–70 000	>5000
G4000SW	13	20 000–7 000 000	4000–500 000	>5000

+ G3000SW + G3000SW and eluted with 0.15 M NaCl solution under the flow rate of 1.0 ml/min. HPAGPLC patterns of reference standards give the distinctly homogeneous shapes with the elution volumes of 35, 42 and 46 ml for LDL, HDL$_2$ and HDL$_3$ by monitoring the absorption at 280 nm as shown in Fig. 1.

   The total lipoprotein fraction containing LDL, HDL$_2$, HDL$_3$ and a small amount of VLDL + chylomicron is obtained by ultracentrifugation in the salt solution of d 1210 at 40 000 rpm for 24 h.

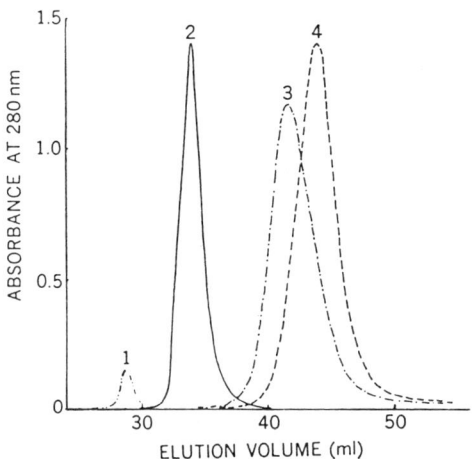

Figure 1. HPLC patterns of standard lipoprotein fractions eluted with 0.1 M NaCl solution. Flow rate: 1.0 ml/min monitoring of the absorption at 280 nm. Peak 1 VLDL (– · · –), 2 LDL (——), 3 HDL$_2$ (– · – ·), 4 HDL$_3$ (– – –).

The total lipoprotein fraction is applied to a similar column system as described in Fig. 1 and the eluent is assayed using phospholipid-P, total cholesterol (TC), and triglyceride (TG), respectively. The elution pattern of total lipoprotein fraction is shown in Fig. 2. The peak of each lipid constituent coincides closely with that of the protein in each lipoprotein fraction designated in Fig. 1. The result shown in Fig. 2 indicates the possible estimation of serum lipoprotein fractions only by quantitation of lipid moieties in using intact whole serum (Okazaki *et al.*, 1980; Ohno *et al.*, 1981).

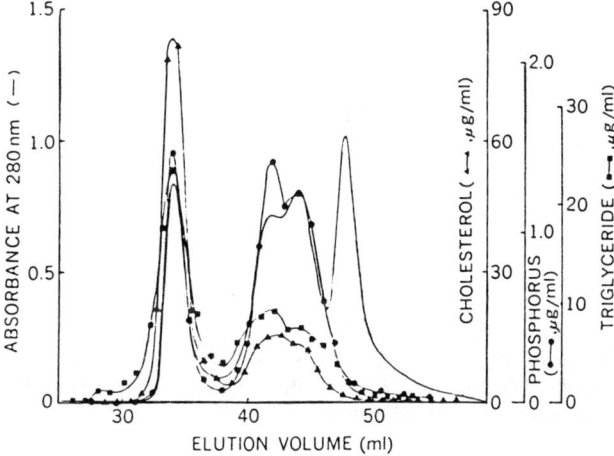

Figure 2. HPLC pattern of total lipoprotein fraction. HPLC is carried out with the same conditions as described in Fig. 1.

For the quantitation of serum lipoproteins by lipid determination the following assay methods were used: TC by Okazaki *et al.* (1980a), TG by Hara *et al.* (1982), and choline-containing phospholipid* by Okazaki *et al.* (1981a), using the enzymatic reagents for routine clinical tests. (The enzymatic reagents were all commercial products.)

The enzymatic reagents are added to the post-column eluent in the reaction tube of chemical derivatization HPLC apparatus (TSK 803 D, Toyo Soda Co., Tokyo).

The different combinations of various kinds of columns described in Table 2 give distinct patterns of serum lipoprotein fractions according to the column characteristics for particle size. The different patterns of the lipoproteins of the same serum by different column systems are shown in Fig. 3.

According to Fig. 3, the column system of G4000SW + G3000SW gives the moderately separated pattern of each lipoprotein fraction. As described later, the column system of G5000PW (G5000PW + G5000PW) is a good tool for VLDL + chylomicron analysis and that of G3000SW + G3000SW + G3000SW is useful for HDL subfraction analysis.

Carrol and Rudel (1983) compared the separation patterns of total lipoprotein fraction with different combinations of column by monitoring the absorption at 280 nm. They concluded the sufficient usefulness of single G5000PW column in the assay of total lipoprotein fraction.

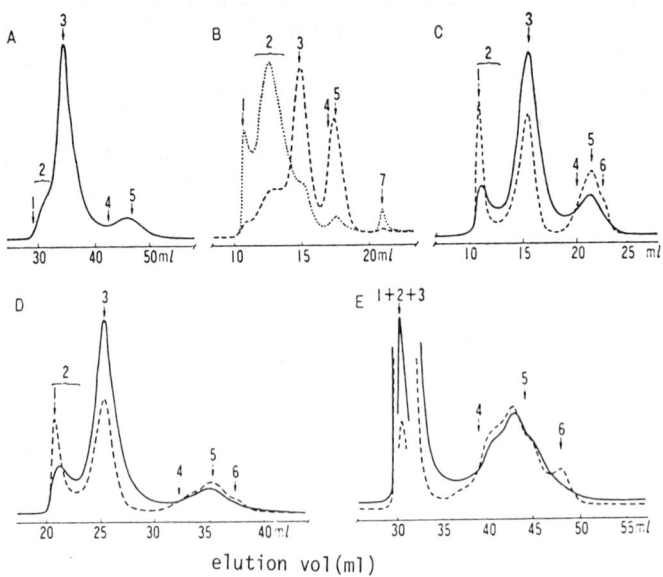

elution vol(ml)

Figure 3. Comparison of HPLC patterns of the same serum by different column systems. (A) TSK GEL G5000PW + G3000SW, (B) G5000PW, (C) G4000SW, (D) G4000SW + G3000SW, (E) G3000SW + G3000SW + G3000SW. Monitoring of TC (——), TG (· · ·) and PL (– – –).

---

* The enzymatic reagent for all phospholipids is not known, but almost all phospholipids in serum lipoprotein are choline-containing phospholipids, phosphatidyl choline, lysophosphatidyl choline, and sphyngomyelin.

Vercaemst *et al.* (1983) reported good separation of each lipoprotein fraction using only a single G4000SW. Carrol and Rudel and Vercaemst *et al.* both examined the total lipoprotein fraction by the absorption at 280 nm.

When the sera of familial hyperlipoproteinemia, which are clinically classified into six groups: I, II_a, II_b, III, IV, and V according to the definition of Fredrickson *et al.* (1967), are applied to the column system of G5000PW + G5000PW and monitored by lipid determination, the patterns of various shapes are obtained as indicated in Fig. 4.

The sera of familial hyperlipoproteinemia present the characteristic feature for each group described by Fredrickson *et al.* The shapes of the patterns drawn by TG, TC, and PL measurements accurately reflect the different contents of each lipid constituent in each lipoprotein fraction.

Among three patterns by lipid analysis, TG may be most suitable to quantitate VLDL + chylomicron, TC may be effective for HDL and LDL regions and choline-containing phospholipid (PL) may be desirable to draw all lipoprotein patterns, because PL distributes, in general, in all lipoprotein fractions.

The column system of TSK GEL G3000SW + G3000SW + G3000SW is used to determine the subfractions of HDL.

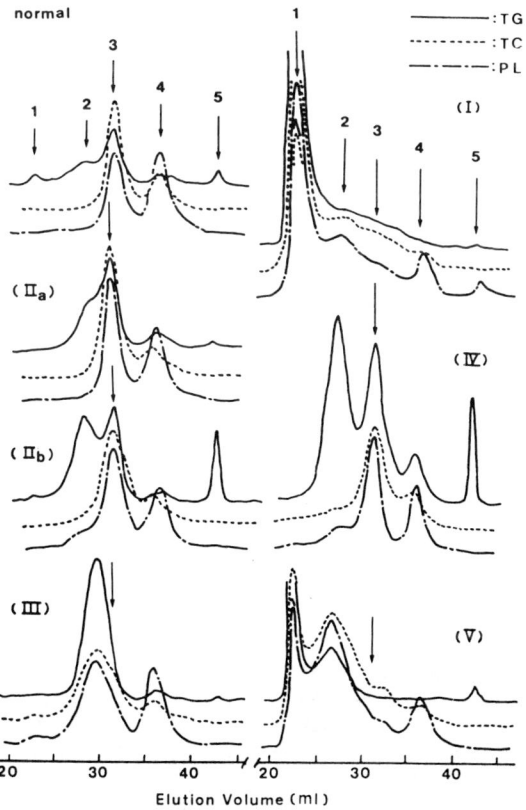

Figure 4. HPLC patterns of familial hyperlipoproteinemia. Column system: G5000SW + G5000PW. (1) chylomicron, (2) VLDL, (3) LDL, (4) HDL, (5) exclusion limit.

The subfractions of HDL have been already isolated with sequential ultra-centrifugation by Anderson *et al.* (1978) and they reported the density ranges of subfractions as follows: HDL_{2b} d 1063–1100; HDL_{2a} d 1100–1125; HDL_3 d 1125–1210, respectively.

Blanche *et al.* (1981) examined HDL fractions prepared by sequential ultra-centrifugation on the gradient gel electrophoresis. By measurement of the relative migration distance (R_F) in gel, two subfractions, $(HDL_{2b})_{gge}$, $(HDL_{2a})_{gge}$ in HDL_2 fraction and three subfractions, $(HDL_{3a})_{gge}$, $(HDL_{3b})_{gge}$, $(HDL_{3c})_{gge}$ in HDL_3 fraction were found.

Several kinds of sera are applied to the column system of G3000SW + G3000SW + G3000SW resulting with HPLC patterns bearing many peaks monitoring of PL as shown in Fig. 5.

For the examination of the relation between those peaks and HDL subfractions, the frequency distribution of the peaks and the shoulders are assayed for 71 sera of normal subjects by Okazaki *et al.* (1982c). The frequency distribution of five groups is shown in Fig. 6.

From the mean elution volumes in Fig. 6 and calibration curve of standard proteins, the particle sizes of five groups are calculated. The particle sizes by HPLC and those by Anderson *et al.* and Blanche *et al.* are shown in Table 3. No discrepancy is found in these results.

By the reason of the broad distribution of PL in all fractions of serum lipo-protein, the relation between HPAGPLC and sequential ultracentrifugation is investigated in assaying PL in the lipoprotein fractions by both methods.

Good correlations are observed in VLDL, LDL, HDL_2, and HDL_3 fractions except in liver disease patients (Okazaki *et al.*, 1983).

Serum lipoprotein contains various kinds of apoproteins indicating the different roles each lipoprotein fraction has in the metabolic pathway.

For the quantitation of apoprotein, Polacek *et al.* (1981) separated apo A-I and A-II and C peptides from the delipidated HDL in 6 M urea solution using TSK

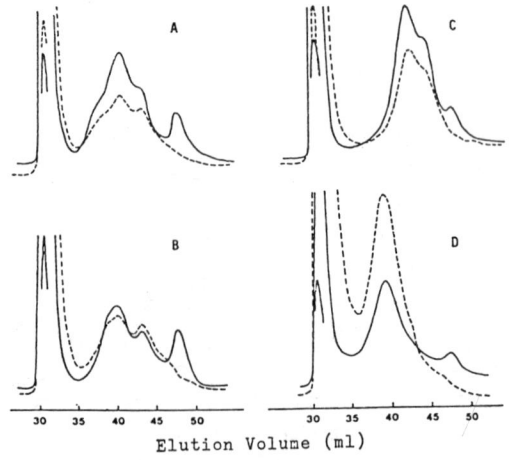

Figure 5. HPLC patterns of HDL fractions of various kinds of sera. Column system: G3000SW + G3000SW + G3000SW. Monitoring of TC (– – –) and PL (———).

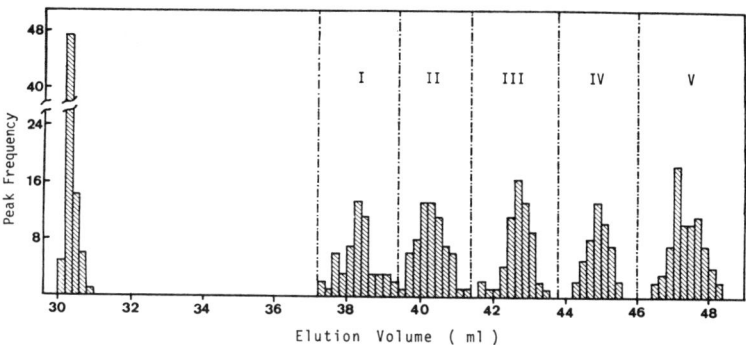

Figure 6. Frequency distribution of peaks in the elution patterns by monitoring of PL for normal subjects (*n* = 71).

Table 3. Particle sizes of HDL subfractions (Å)

Okazaki *et al.* (1982)	Anderson *et al.* (1978)	Blanche *et al.* (1981)
Group I 122.0 ± 2.8	HDL$_{2b}$ 108–120	(HDL$_{2b}$)$_{gge}$ 105.7
Group II 110.1 ± 2.1	HDL$_{2b}$ 97–107	(HDL$_{2a}$)$_{gge}$ 91.6
	HDL$_3$ 85–96	(HDL$_{3a}$)$_{gge}$ 84.4
Group III 97.5 ± 1.8		(HDL$_{3b}$)$_{gge}$ 79.7
Group IV 86.7 ± 1.3		(HDL$_{3c}$)$_{gge}$ 76.2
Group V 76.3 ± 1.6		

GEL G3000SW. Pfaffinger *et al.* (1983) analyzed apo VLDL in 4 M guanidine HCl solution using Bio-Gel TSL Guard column + Bio-Gel TSK 50 + Bio-Gel TSK 400 + Spherol TSK 300 and they could identified the peaks of apo B, apo C, and apo E, respectively.

Socorro *et al.* (1982) tried to separate apo LDL using TSK GEL G3000SW resulting with the decomposed products.

On the reversed phase HPLC, Ronan *et al.* (1982) isolated the homogeneous apo C-II from partially purified C-II by DEAE cellulose chromatography. They used Radial Pak C18 cartridge with linear gradient of 0.01 M ammonium bicarbonate and acetonitrile to give a good separation of apo C-III$_2$, C-III$_1$, C-III$_3$, and C-II in this order.

Ott and Shore (1982) isolated apo A-I$_1$, A-I$_2$, and A-II from apo HDL with linear gradient of Tris HCl in 6 M urea at pH 7.9 using SynChropak AX 300 and they also obtained apo C-I, C-II, C-III$_1$, and C-III$_2$ peaks from apo VLDL using identical conditions.

These apoprotein analyses started from delipidated lipoprotein fractions, apo HDL, or apo VLDL, but the delipidation procedure is troublesome and difficult to some extent. Kinoshita *et al.* (1983, 1984) succeeded to assay apo A-I and A-II from HDL in phosphate buffer solution containing 0.1% SDS only by heating this solution at 60°C for 5 min as shown in Fig. 7. These apoprotein peaks coincide exactly with those obtained from delipidated HDL by a similar procedure.

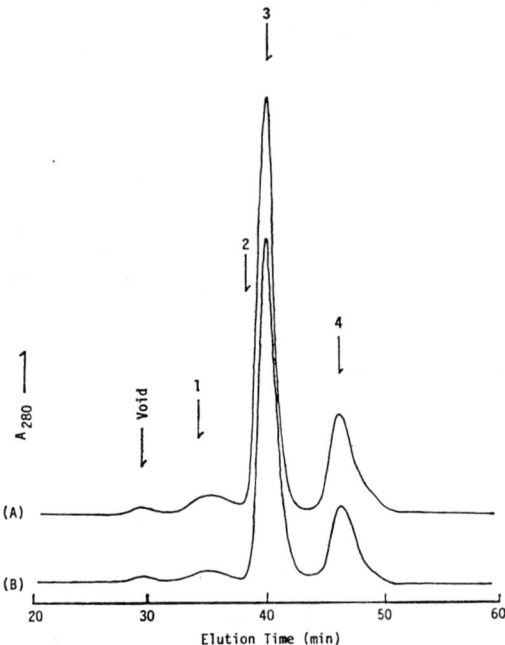

Figure 7. HPLC patterns of HDL apoproteins. Eluted with 0.1 M phosphate buffer solution pH 7.0 containing 0.1% SDS. Column system: G3000SW at the flow rate of 0.33 ml/min. (A) HDL fraction, (B) apo HDL fraction. (1) albumin, (2) apo E, (3) apo A-I, (4) apo A-II.

CONCLUSIONS

From these experimental results, high-performance aqueous gel permeation liquid chromatography is the most promising method for the quantitation of serum lipoprotein, because this is performed in less than 1 h with only 5–20 μl of intact serum.

REFERENCES

Anderson, D. W., Nichols, A. V., Pan, S. S., and Lidgren, F. T. (1978). High density lipoprotein distribution, resolution and determination of three major components in a normal population sample. *Atherosrelosis 29*, 161–179.

Blanche, P. J., Gong, E. L., Forte, T. M., and Nichols, N. V. (1981). Characterization of human serum high density lipoproteins by gradient gel electrophoresis. *Biophys. Biochim. Acta 665*, 308–419.

Carrol, R. M. and Rudel, L. L. (1983). Lipoprotein separation and low density lipoprotein molecular weight determination using high performance liquid gel filtration chromatography. *J. Lip. Res. 24*, 200–207.

de Lalla, O. F. and Gofman, J. W. (1954). Ultracentrifugal analysis of serum lipoproteins. *Method of Biochemical Analysis 1*, 459–478.

Fredrickson, D. S., Levy, R. T., and Lee, R. S. (1967). Fat transport in lipoprotein — an integrated approach to mechanism and disorders. *New Eng. J. Med. 276*, 34–42; 94–102; 148–156; 215–225; 273–281.

Hara, I., Shiraishi, K., and Okazaki, M. (1981). High performance liquid chromatography of human serum lipoproteins. Selective detection of triglycerides by enzymatic reaction. *J. Chromatogr. 239*, 549–557.

Havel, R. J., Eder, H. A., and Bragdon, J. H. (1955). The distribution and chemical compositions of ultracentrifugally separated lipoproteins in human serum. *J. Clin. Invest. 34*, 1345–1353.

Kinoshita, M., Okazaki, M., Kato, H., Teramoto, T., Matsushima, T., Naito, C., Oka, H., and Hara, I. (1983). Simple method for analysis of apolipoprotein in HDL by high performance liquid chromatography. *J. Biochem. 94*, 615–617.

Kinoshita, M., Okazaki, M., Kato, H., Teramoto, T., Matsushima, T., Naito, C., Oka, H., and Hara, I. (1984). Analysis of apolipoproteins in HDL by high performance liquid chromatography. *J. Biochem. 95*, 1111–1118.

Kostner, G. M. and Holasek, A. (1977). The separation of human serum high density lipoproteins by hydroxyapatite column chromatography. *Biophys. Biochem. Acta 488*, 417–431.

Ohno, Y., Okazaki, M., and Hara, I. (1981). Fractionation of human serum lipoproteins by high performance liquid chromatography I. *J. Biochem. 89*, 1675–1680.

Okazaki, M., Ohno, Y., and Hara, I. (1980). High performance aqueous gel permeation chromatography of human serum lipoprotein. *J. Chromatogr. 221*, 257–264.

Okazaki, M., Ohno, Y., and Hara, I. (1981a). Rapid method for the quantitation of cholesterol in human serum by high performance liquid chromatography. *J. Biochem. 89*, 879–887.

Okazai, Y., Shiraishi, K., Ohno, Y., and Hara, I. (1981b). High performance aqueous gel permeation chromatography of serum lipoproteins: Selective detection of cholesterol by enzymatic reaction. *J. Chromatogr. 223*, 285–293.

Okazaki, M., Hagiwara, N., and Hara, I. (1982a). Quantitation method of choline-containing phospholipids in human serum lipoproteins by high performance liquid chromatography. *J. Biochem. 91*, 1381–1399.

Okazaki, M., Hagiwara, N., and Hara, I. (1982b). High performance liquid chromatography of human serum lipoproteins. Selective detection of choline-containing phospholipids by enzymatic reaction. *J. Chromatogr. 231*, 13–23.

Okazaki, M., Hagiwara, N., and Hara, I. (1982c). Heterogeneity of human serum lipoproteins on high performance liquid chromatography. *J. Biochem. 92*, 517–524.

Okazaki, M., Itakura, H., Shiraishi, K., and Hara, I. (1983). Serum lipoprotein measurement-Liquid chromatography and sequential floatation compared. *Clin. Chem. 29*, 768–773.

Ott, G. S. and Shore, V. G. (1982). Anion-exchange high performance liquid chromatography of human serum lipoproteins. *J. Chromatogr. 231*, 1–12.

Pfaffinger, D., Edelstein, C., and Scanu, A. M. (1983). Rapid isolation of apolipoprotein E from human plasma VLDL by molecular sieve high performance liquid chromatography. *J. Lip. Res. 24*, 796–800.

Polacek, D., Edelstein, C., and Scanu, A. M. (1981). Rapid fractionaion of human high density apolipoproteins by high performance liquid chromatography. *Lipids 16*, 927–929.

Ronan, R., Kay, L. L., Martha, K., Meng, S., and Brewer, Jr H. B. (1982). Purification and characterization of apolipoprotein C-II from human plasma by high performance liquid chromatography. *Biophys. Biochim. Acta 713*, 657–662.

Sata, T., Estrich, D. C., Wood, P. D., and Kensell, L. W. (1970). Evaluation of gel chromatography of plasma lipoprotein fraction. *J. Lip. Res. 11*, 331–340.

Socorro, L., Lopez, F., Lopez, A., and Camejo, G. (1982). ApoLDL: evidence for aggregating system of heterogeneous subunits. *J. Lip. Res. 23*, 1283–1291.

Vercaemst, R., Rosseneu, M., and Bierliet, J. P. (1983). Separation and quantitation of plasma lipoproteins by high performance liquid chromatography. *J. Chromatogr. 276*, 174–181.

Progress in HPLC, Vol. 1, pp. 105–131
Parvez *et al.*, (Eds)
© 1985 VNU Science Press

Heterogeneity of renin: application of HPLC in studies on renin and renin binding protein

HIROSHI IWAO,[1] SHOKEI KIM,[1] YOSHIHIKO FUNAE,[2] NORIFUMI NAKAMURA,[1] FUMIHIKO IKEMOTO,[1] and KENJIRO YAMAMOTO [1]*

[1] Department of Pharmacology and [2] Laboratory of Chemistry, Osaka City University Medical School, Asahimachi, Abeno-ku, Osaka 545, Japan

INTRODUCTION

Chromatography is widely used in studies of biochemistry, including gel permeation, ion exchange, gas, thin-layer, paper, and affinity chromatography (Regnier and Gooding *et al.*, 1980). However, the carbohydrate and polyacrylamide gel materials used as column supports have a serious disadvantage in that they are mechanically weak, therefore, conventional chromatographic resolution of proteins involves a long process under low pressure. Ion-exchange and gel permeation chromatography can now be done under high pressure, because a supporting column of mechanical strength has been designed. With this development of high-performance liquid chromatography (HPLC), on which the elution time for the fractionation of proteins is surprisingly shorter than that on classical chromatography, HPLC separations may become routine (Regnier, 1982; Vanecek and Kegnier, 1980; Varik and Taen, 1982; Chang *et al.*, 1976).

Renin is an aspartyl protease mainly synthesized in juxtaglomerular cells, located in the glomerular afferent arteriole in the kidney and released into the blood stream, (Edelman and Hartroft, 1961; Tobian *et al.*, 1959). Within the plasma, renin hydrolyzes angiotensinogen very specifically to release a decapeptide angiotensin I which has no direct depressor action (Braun-Menendez *et al.*, 1940; Page and Helmer, 1940). Subsequently when angiotensin I passes through capillary beds, particularly the lungs, it is converted by angiotensin converting enzyme (Nakajima *et al.*, 1973) to an octapeptide angiotensin II. This is a most potent pressor substance, has vasoconstrictive actions, and stimulates the secretion of the sodium retaining hormone aldosterone in the adrenal gland (Campbell *et al.*, 1974). Thus, as the formation of angiotensin I by renin is the initial and rate-limiting step of the renin-angiotensin-aldosterone cascade, renin is considered to play an important role in the regulation of blood pressure and in sodium and body fluid homeostasis (Skeggs *et al.*, 1980; Haber and Carlson *et al.*, 1982).

---

* To whom correspondence should be addressed.

Despite numerous attempts to purify this enzyme from the kidney, the complete isolation of pure and stable renal renin from the kidney proved to be most difficult due to its extremely low concentration and rapid inactivation by co-eluted proteases (Hass *et al.*, 1953; Peart *et al.*, 1966; Waldhausl *et al.*, 1970). Success awaited the development of affinity chromatography, using renin inhibitors. Pepstatin is an extremely potent inhibitor of renin isolated by Umezawa (1973) from *Streptomyces* strains (Murakami *et al.*, 1973; Corvol *et al.*, 1973; Aoyagi *et al.*, 1972). Renin was isolated from hogs (Murakami and Inagami, 1975; Inagami and Murakami, 1977; Corvol *et al.*, 1977), dogs (Dzau *et al.*, 1979), rats (Matoba *et al.*, 1978), and from humans (Galen *et al.*, 1979; Yokosawa *et al.*, 1980; Slater and Strout, 1981; Higaki *et al.*, 1982). However, it is very difficult to obtain these renin preparations in sufficient quantities for full characterization, such as the determination of the primary structure and the active site, except for juxtaglomerular cell tumor. For example, only 1.3 mg of human pure renal renin was obtained from 35 kg of human kidney (Higaki *et al.*, 1982), 1.86 mg from about 5 kg of rat kidney (Matoba *et al.*, 1978), 2.3 mg from 47 kg of hog kidney (Corvol *et al.*, 1977). Therefore, the purification of renal renin is difficult and application of renal renin on HPLC of gel permeation and ion-exchange chromatography is currently not so feasible.

The submaxillary glands of adult male mice contain much larger amounts of renin than in the kidney (Bing and Poulsen, 1971). Mouse submaxillary renin was first purified by Cohen *et al.* (1972) using five steps of conventional chromatography and rapid and large-scale purifications were then developed by Suzuki *et al.* (1981), Ho *et al.* (1982) and Misono *et al.* (1982a). Submaxillary gland renin is the only renin preparation available in large amounts. Although its physiological function has not been completely elucidated, it is known to have physiochemical, enzymatic, and immunological properties similar to renal renin (Michelakis *et al.*, 1974a, b; Bing *et al.*, 1980). Therefore, numerous studies using this renin as an experimental model of renal renin have been done (Corvol, 1983a), including the characterization of the active site by Misono and Inagami (1980), determination of the primary structure of this renin by Misono *et al.* (1982a, b) and of its precursor by Panthier *et al.* (1982), the demonstration of the biosynthesis of a prorenin and its processing by Poulsen *et al.* (1979); Rougeon *et al.* (1981); Catanzaro *et al.* (1983); Corvol *et al.* (1983a), and studies of the metabolism in the kidney (Iwao *et al.*, 1982b, c, 1983a, b). We also used submaxillary renin as a model of renal renin for studies using high performance ion-exchange chromatography.

One of the important and interesting characteristics of renin is its heterogeneity. Generally, renin has a molecular weight of around 40 000, but renin with a molecular weight over 40 000 has been found in plasma, amniotic fluid, and kidney in many species. The significance of this high molecular weight renin is unknown. Some of the high molecular weight renins are a complex of low molecular weight renin and renin binding substance (Boyd, 1974; Leckie and McConnel, 1975; Inagami *et al.*, 1977; Funakawa *et al.*, 1978; Kawamura *et al.*, 1979; Takaori *et al.*, 1981; Ikemoto *et al.*, 1982). Inagami *et al.* (1977) reported the conversion of low-molecular weight renin to high molecular weight renin in the presence of sodium tetrathionate. In our laboratory, high molecular weight renin and renin binding protein have been extensively investigated and high performance gel permeation chromatography on G3000SW was an important tool in these studies.

MATERIALS AND METHODS

Packed column

Cation-exchange columns including IEX535CM and SP-5PW, and anion-exchange columns including DEAE-3SW and DEAE-5PW were used in our studies of high-performance ion-exchange chromatography (HPIEC) of mouse submaxillary gland renin. The high-performance gel permeation chromatography (HPGPC) was performed on G3000SW. All of these columns were supplied in a prepacked form by Toyo Soda Co., Yamaguchi, Japan.

IEX535CM and DEAE-3SW are derivatives of TSK-GEL G3000SW, have a pore size of *ca.* 250 Å, and use macroporous spherical silica with a particle diameter of *ca.* 10 μm. which is chemically weak and cannot be used over pH 8.0.

On the other hand, SP-5PW and DEAE-5PW are derivatives of TSK-GEL G5000PW with a pore size of *ca.* 1000 Å, and their base materials are not silica but rather chemically stable hydrophilic resins which can be used at a broad range of pH. Characteristics of these ion-exchange columns are listed in Table 1.

Table 1. Characteristics of the ion-exchange columns

	Functional group	Formula	Base-materials	Particle size (μm)	Pore size (Å)	Characteristics
IEX-535CM	Carboxy-methyl	$-CH_2CO_2-$	silica	10	250	Weak cation exchanger
DEAE-3SW	Diethylamino-ethyl	$-C_2H_4N^+H(C_2H_5)_2$	silica	10	250	Weak anion exchanger
SP-5PW	Sulfonic-propyl	$-C_3H_6SO_3-$	polymer	10	1000	Strong cation exchanger
DEAE-5PW	Diethylamino-ethyl	$-C_2H_4N^+H(C_2H_5)_2$	polymer	10	1000	Strong anion exchanger

Apparatus

Apparatus for HPIEC. The pumping system for the gradient elution by high-performance ion-exchange chromatography consisted of a SP 8750 organizer (Toyo Soda) and an SP 8700 solvent delivery system (Toyo Soda). The detector was either a UV-8 (Toyo Soda) or a spectrophotometric detector SPD-2A (Shimadzu, Kyoto, Japan). The recorder was obtained from Toyo Soda. Sample injection was achieved with a Model 7125 syringe loading sample injector (Rheodyne, Berkeley, Calif.) with a 500 μl loop. Microsyringes were purchased from Hamilton Co. (Reno, Nev.). Single Fracon SF-60L (Toyo Kagaku Sangyo Co.) was used as a fraction collector.

Apparatus for HPGPC (High-performance gel permeation chromatography). The high pressure pump was supplied by Hitachi, Tokyo Japan and a precolumn, TSK GSWP (from Toyo Soda) was used to protect the G3000SW column. The recorder was the product of Hitachi. Other equipment was the same as used for ion-exchange chromatography.

Renin activity

Renin activity was determined by modifying the angiotensin I radioimmunoassay method (Haber *et al.*, 1969; Stockigt *et al.*, 1971; Menard and Catt, 1972). Plasma from nephrectomized rats was used as the renin substrate. The animals were bilaterally nephrectomized and 20–24 h later blood samples were obtained. The plasma was separated by centrifugation at 4°C and then used as the renin substrate. The titer of angiotensinogen was 8000 ng of angiotensin I liberated from 1 ml of the plasma.

One ml of the incubation mixture consisted of 50 μl of sample, 100 μl of nephrectomized rat plasma, 850 μl of 0.25 M sodium phosphate buffer pH 7.4, 2 mM disodium ethylenediamine tetracetate (EDTA, Wako Chemicals), 2 mM phenylmethylsulfonylfluoride (Sigma Chemical Co.), 2 mM 8-hydroxyquinoline sulfate (Ishizu Pharmaceutical Co.), and 100 μM diisopropylfluoro phosphate (Fluka AG Chemical). All four angiotensinase and protease inhibitors have to be included in the incubation mixture, because only in their presence was degradation of angiotensin I during incubation prevented, since extracts from kidney and submaxillary gland contained significant amounts of angiotensinase activity. These inhibitors were found to have no effect on angiotensin I radioimmunoassay (the data are not shown). The enzymatic reaction was carried out at 37°C for 30 min and terminated by cooling.

[125 I] Angiotensin I (5-L-isoleucine), specific activity 2200 ci/mmol was obtained from New England Nuclear. Antibodies were elicited in the rabbit using ileu[5]-angiotensin I, using the method of Freedlender *et al.* (1974). The cross-reactions of the antibody against ileu[5]-angiotensin II, ileu[5]-angiotensin III, and rat tetradecapeptide were 0.01%, 0.13%, and 0.13% respectively, and against Sar[1], Ala[8]-angiotensin II, and nephrectomized rat plasma (8000 ng of angiotensin I/ml) were hardly detectable below 100 μg/ml and 2000 ng of angiotensin I/ml, respectively. The sensitivity of the assay was 0.05 ng/ml. In preparing incubation mixtures for angiotensin I radioimmunoassay, 100 μl of [125 I] angiotensin I containing about 15 000 cpm and 100 μl of 1:3000 angiotensin I antibody were added to 700 μl of 40 mM sodium phosphate buffer (pH 7.4), containing 0.1% bovine serum albumin (BSA, Seikagaku, Kogyo) and 1 mM EDTA and either 100 μl of standard solution of angiotensin I (0.05–0.8 ng/ml) or incubated sample. The incubation was allowed to proceed for 20 h at 4°C. After incubation, 500 μl of dextran-coated charcoal solution was added to the mixture to separate antibody bound and free angiotensin I. The dextran-coated charcoal suspension consists of 10 g of Norit SX-3 (American Norit), 1 g of dextran T70 (Pharmacia Fine Chemicals), and 660 ml of 40 mM sodium phosphate buffer (pH 7.4) containing 0.1% BSA and 1 mM EDTA. After standing for 10 min at 4°C, the preparations were centrifuged at 3000 rpm for 10 min at 4°C, and the radioactive precipitate counted (as free phase) in an auto-gamma scintillation counter (Packard, 800c). The count of non-specific binding was subtracted from the count of each sample.

Direct radioimmunoassay of mouse renin

For the measurement of absolute amounts of renin, direct radioimmunoassay of mouse renin was performed using slight modifications to the method of Michelakis *et al.* (1974b).

Mouse submaxillary renin A was used for the preparation of [^{125}I] renin. Purified renin A (Cohen *et al.*, 1972) was labeled with ^{125}I, by the chloramine-T method (Michelakis *et al.*, 1974b). The separation of the labeled renin was carried out on a column of Sephadex G-25.

Anti-renin anti-serum was newly elicited in rabbits. The titer of anti-serum was tested for its ability to produce 50% binding of [^{125}I] renin, which was measured by incubation of [^{125}I] renin, (around 6000 cpm/100 μl) with various dilutions of anti-serum (1:10^2 to 1:5 $\times$ 10^5). The titer of anti-serum was 1:3 $\times$ 10^5 in final dilution (Michelakis *et al.*, 1974b; Iwao *et al.*, 1980).

One hundred μl of standard of renin A, or a certain amount of each unknown sample was mixed with 100 μl of renin antiserum (1:30 000), and then 100 μl of [^{125}I] renin (6000 cpm) was added. The final volume of each tube was adjusted to 1.0 ml by 50 mM, phosphate buffer, pH 7.4, containing 0.5% bovine serum albumin (BSA, Miles Laboratories) The tubes were then incubated for three days at 4°C. At the end of the incubation, 100 μl of 1% γ-globulin (Sigma Chemical) was added, the preparation mixed, and then 1.0 ml of 20% of polyethylene glycol (mol. wt 6000, Fluka) was added. After being left to stand for 5 min, the preparations were centrifuged at 3000 rpm for 30 min. The supernatant and precipitate were counted in an auto-gamma scintillation counter.

Protein assay

Protein contents were determined by the methods of Lowry *et al.* (1951), and Bensadoun and Weinstein (1976), using bovine serum albumin (Seikagaku Kogyo Co.) as a standard.

Polyacrylamide gel electrophoresis

Polyacrylamide gel electrophoresis without sodium dodecyl sulfate (SDS, Wako Chemicals) was run in 7% polyacrylamide gels using the method of Davis (1964). The electrophoresis was carried out using a constant current of 2 mA/gel.

SDS-polyacrylamide gel electrophoresis

Polyacrylamide gel electrophoresis with SDS was run in 12% polyacrylamide gels, according to the method of Laemmli (1970). Samples dissolved in the stacking gel buffer were heated to 100°C in water for 2 min in the presence of 2-mercapto-ethanol (Nakarai Co). Protein standards (Bio-Rad Laboratories) were used including lysozyme, soybean trypsin inhibitor, carbonic anhydrase, ovalbumin, bovine serum albumin, and phosphorylase B. Protein bands in the gel were stained with 0.1% Coomassie brilliant blue R-250 (Eastman Kodak Co., Rochester, N.Y.).

Preparation of pepstatin-aminohexyl sepharose
Pepstatin A was obtained from the Protein Research Foundation (Mino, Osaka). Aminohexyl sepharose 4B was purchased from Pharmacia Fine Chemicals. Pepstatin A was coupled to aminohexyl sepharose by the method of Murakami and Inagami (1975). One ml of the wet gel contained approximately 0.5 μmol of covalently bound pepstatin, determined by amino acid analysis.

Purification of renin from mouse submaxillary gland
Mouse submaxillary renin was purified by two different methods as follows:

(1) Purification by the method of Cohen *et al.* (1972).

Extraction procedure. Adult male albino mice of the ddY strain were anesthetized with pentobarbital and the submaxillary glands excised and frozen until use. About 24 g of frozen tissues were homogenized with 100 ml of cold distilled water at 4°C, and the preparation centrifuged for 10 min at 16 000g. The precipitate was re-homogenized with 90 ml of cold distilled water followed by re-centrifugation. The two supernatants were combined and treated with streptomycin sulfate (0.75%) at pH 7.0 for 12 h to remove mucin. After centrifugation, solid ammonium sulfate was added to the supernatant (56 g per 100 ml of the supernatant), the mixture centrifuged for 10 min at 16 000g at 3°C, and the residue suspended in cold distilled water. The suspension was dialyzed against distilled water overnight and then centrifuged.

Step 1. Sephadex G100. The supernatant was concentrated to 40 ml by ultra-filtration with an Amicon YM10 membrane. Half as much as the preparation (20 ml) was applied to a column of Sephadex G100 (970 X 50 mm) previously equilibrated with 10 mM Na-acetate buffer, pH 5.9, containing 0.1 M NaCl, while the remainder was stored until the next experiment.
The fractions containing renin eluted with the same buffer were pooled, concentrated, adjusted to pH 4.5 by ultrafiltration with 0.01 M Na-acetate buffer, pH 4.5, and centrifuged at 3000 rpm for 10 min.

Step 2. DE-cellulose. The supernatant was applied to a column of DE-cellulose (120 X 15 mm, Whatman DE52) pre-equilibrated with the same buffer. The eluate containing renin and which passed through the column, was concentrated by ultrafiltration, adjusted to pH 5.4, and dialyzed against 0.05 M sodium acetate, pH 5.4.

Step 3. CM-cellulose. The dialyzed preparation was applied to a column of CM-cellulose (150 X 15 mm, Whatman CM52). Renin fractions were eluted with a linear gradient of NaCl (0—0.15 M) in 0.05 M sodium acetate, pH 5.4, concentrated and adjusted to pH 7.5 with 0.02 M Tris-HCl buffer and then dialyzed against the same buffer.

Step 4. DE-cellulose. The dialyzed sample was applied to DE-cellulose (150 X 15 mm). The fractions with renin activity eluted with a NaCl gradient (0—0.3 M) in 0.02 M Tris-HCl buffer were pooled, concentrated, and adjusted to pH 5.4.

Step 5. Final CM-cellulose. The above concentrated solution was chromato-graphed on a CM-cellulose column (150 × 15 mm) pre-equilibrated with 0.05 M sodium acetate, pH 5.4, and renin activity was eluted with the same linear gradient of NaCl as that of the previous CM-cellulose chromatography (step 3).

(2) Purification using IEX535CM.

Renin from mouse submaxillary gland was purified by a novel method, that is a two-step procedure including chromatography on a pepstatin aminohexyl-sepharose column (Murakami and Inagami, 1975, Devaux *et al.*, 1976; Suzuki *et al.*, 1981) and high-performance liquid chromatography on a IEX535CM column. In brief, submaxillary glands (7 g) from adult male albino mice of ICR strain were homogenized in 70 ml of 0.02 M sodium acetate buffer, pH 5.5, without protease inhibitor, at 4°C. The homogenate was centrifuged at 100 000g for 60 min. The supernatant was applied to a column of pepstatin-aminohexyl-sepharose-4B (90 × 10 mm) previously equilibrated with 0.02 M sodium acetate buffer, pH 5.5. The column was washed with the same buffer and eluted with 0.02 M sodium acetate buffer, pH 5.5 containing 0.5 M NaCl, 0.1 M Tris-HCl buffer, pH 7.4, and then 0.1 M Tris-HCl, pH 7.4 containing 0.5 M NaCl. The fractions including renin, which were eluted with 0.1 M Tris-HCl pH 7.4 containing 0.5 M NaCl, were pooled, adjusted to pH 5.4 with 100 ml of 0.05 M sodium acetate, pH 5.4, and concentrated to 2 ml by ultrafiltration with an Amicon YM10 membrane. The concentrated preparation was applied to a column of IEX535CM pre-equilibrated with 0.05 M sodium acetate buffer, pH 5.4. Protein peaks were monitored on absorbance at 280 nm and collected for determination of protein and renin activity. Renin peaks were then applied to gel electrophoresis.

RESULTS AND DISCUSSION

High-performance gel permeation chromatography on G3000SW

High-speed gel filtration chromatography of renin was performed on TSK G3000SW, a microparticulate silica gel chemically bonded with hydrophilic compounds, and with little adsorption for proteins (Fukano *et al.*, 1978) this may be the most useful of TSK-GEL SW columns for the separation of proteins as it exhibits the highest separation efficiency for a molecular weight of 30 000–500 000 (Kato *et al.*, 1980a, b).

Figure 1 shows the HPLC profile of renin A and standard proteins on G3000SW (600 × 7.5 mm i.d. × 2). Renin A and standard proteins could be completely separated, and renin A showed a symmetric single peak with a retention time of 36.5 min and a molecular weight of 38 000. Recoveries of BSA and renin were over 90% and then decreased gradually, after repeated use the repeated application of plasma decreased the recovery to around 60% (after about 100 applications). The reproducibility of the chromatogram was excellent with about 50 applications of plasma, often the retention time was gradually delayed and a symmetrical profiles of u.v. 280 nm absorbance were maintained. Thus, HPLC of G3000SW is superior to classical chromatography using Sephadex, with regard to sharpness of separation, the time required, and reproducibility.

Several examples using G3000SW will be given below.

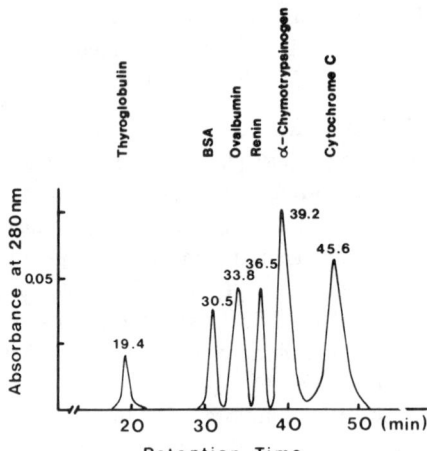

Figure 1. High-performance gel permeation chromatographic profile of standard proteins and renin A on a G3000SW column. Thyroglobulin, bovine serum albumin (BSA), ovalbumin, α-chymotrypsinogen, and cytochrome *c* were used as standard proteins. The sample was applied to double column system of TSK G3000SW (600 × 7.5 mm × 2) equilibrated with 50 mM sodium phosphate buffer (pH 7.4) containing 0.1 M NaCl. The flow rate was 1 ml/min and the temperature was approximately 20°C. The retention time of each protein is shown on the top of the peak. In this double column system, renin A was eluted at 36.5 min.

Application of high molecular weight renin and renin binding substance to GPC on G3000SW. One of the important characteristics of renin is its heterogeneity: renin has multiple forms with different molecular weights and studies concerning the difference in molecular weight are numerous. Renin is commonly thought to have a molecular weight of around 40 000: low molecular weight renin (l.m.w. renin). Renin over this molecular weight is generally called high molecular weight renin (h.m.w. renin). High molecular weight renin has been detected in plasma, amniotic fluid, and the kidney of various species (Day and Leutscher, 1974; Day *et al.*, 1975; Murakami *et al.*, 1980). Boyd (1974) first reported that h.m.w. renin was present in hog kidney and consisted of l.m.w. renin and renin binding substance. Later, h.m.w. renin was found to be a complex of l.m.w. renin and renin binding substance, in the dog (Funakawa *et al.*, 1978; Kawamura *et al.*, 1979), hog (Inagami *et al.*, 1977), and rat kidney (Sagnella *et al.*, 1980; Takaori *et al.*, 1981). We used HPLC on G3000SW to study characteristics of h.m.w. renin and renin binding substance.

Iwao *et al.* (1982a) identified h.m.w. renin and renin binding substance in mouse kidney using HPLC on G3000SW (600 × 7.5 mm i.d. × 2). Figure 2 (a and b) shows HPLC profiles of renin prepared from the isolated renin granule of mouse kidney. Renin activity in the renin granule without sodium tetrathionate, a potent oxidant for sulfhydryls, showed a single peak with a retention time of 36.5 min and a molecular weight of approximately 38 000, such being the same as renin A (Fig. 2a). In the presence of sodium tetrathionate, the retention time of renin activity was exactly the same as the above case (Fig. 2b). These results indicate that

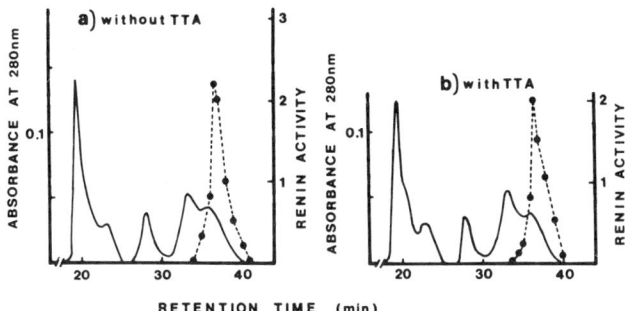

Figure 2. HPLC patterns of renin prepared from mouse renal renin granule fraction on the above G3000SW double column system, (a) without sodium tetrathionate (TTA) and (b) with TTA. The solid line indicates the absorbance at 280 nm and closed circles and broken line indicate renin activity (μg angiotensin I/ml/h). The elution conditions were the same as in Fig. 1. Renin activity eluted at the same retention time as in Fig. 1.

the renin binding substance was not present in the renin granules, therefore even with sodium tetrathionate, l.m.w. renin could not be converted to h.m.w. renin.

When mouse renal cortical supernatant fraction without sodium tetrathionate was applied to a column of G3000SW, the renin activity showed a single peak with the retention time of 36.5 min, the same as in renin granules (Fig. 3a). However, renin activity in the supernatant with sodium tetrathionate showed two peaks with a retention time of 32 and 36.5 min (Fig. 3b). To confirm whether both activities were indeed true renin, a specific renin antibody against mouse sub-maxillary renin was used. Both activities were completely neutralized by a specific antiserum, thereby indicating that both activities are true renin and immunologically identical.

To determine the absolute amount of renin, direct radioimmunoassay was used. As shown in Fig. 4, the relationship between the enzymatic activity and absolute

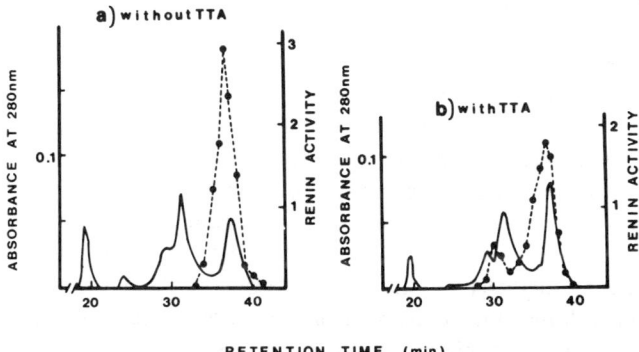

Figure 3. HPLC profiles of renal cortical supernatant on G3000SW double column system, (a) without sodium tetrathionate (TTA) and (b) with TTA. The solid line indicates the absorbance at 280 nm and solid circles and dashed line indicate renin activity (μg angiotensin I/ml/hr). The elution conditions were the same as in Fig. 1.

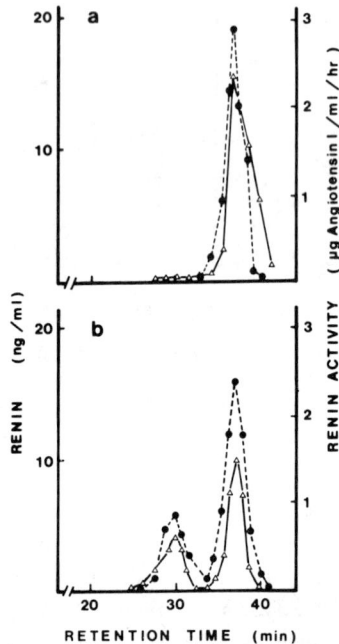

Figure 4. HPLC pattern of renal supernatant on G3000SW double columns system, (a) without sodium tetrathionate (TTA) and (b) with TTA. The dashed line and solid circles indicate renin activity. The solid line and open triangles indicate absolute amount of renin determined by direct radioimmunoassay. The elution conditions were the same as in Fig. 1.

amount of h.m.w. renin from the kidney supernatant was identical to that of l.m.w. renin.

Thus, renin in the renin granule was found to be stored in a low molecular weight form of around 38 000. We confirmed that renin binding substance existed in the supernatant fraction of mouse renal cortex, that l.m.w. renin bound with renin binding substance and subsequently converted to h.m.w. renin in the presence of sodium tetrothionate.

Ikemoto *et al.* (1982) determined the exact localization of renin binding substance in the renal cortex. TSK G3000SW column was also successfully used. Glomeruli containing juxtaglomerular cells and tubular segments were isolated from the rat kidney, using the method of Imbelt-Teboul *et al.* (1978). In this experiment, the contents of protein and renin were extremely small. The renin activities were about 300 pmol angiotensin I/h/100 μl in the extract of glomeruli but were hardly detected in the tubular segment, thereby indicating that the isolated glomeruli but not the isolated tubular segments contained juxtaglomerular cells. This was confirmed microscopically.

The mixture containing 45 μl of the extract of glomeruli and 5 μl of 50 mM sodium tetrathionate was incubated at 15°C for 20 min and then applied to G3000SW. As shown in Fig. 5(a), despite the presence of sodium tetrathionate, only a single peak of the renin activity was observed at a retention time of 36.5

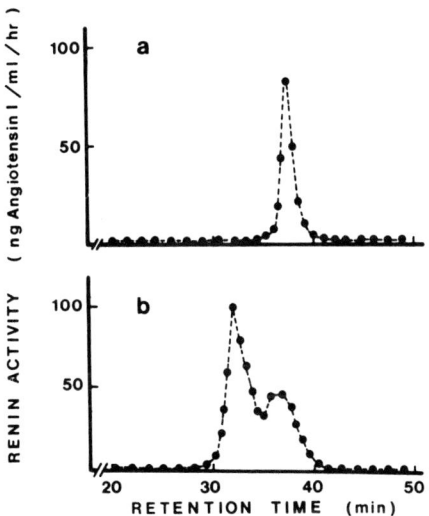

Figure 5. HPLC profile of renin in the extract of isolated glomeruli containing afferent arterioles (a), and the mixture of l.m.w. renin and the extract of renal cortical tubular segments (b) on two coupled G3000SW columns. In both cases, sodium tetrathionate (TTA) was added. The elution conditions were the same as in Fig. 1.

min, that is identical to that of l.m.w. renin prepared from renin granules. These findings indicate that the renin binding substance was absent in the extract of the isolated glomeruli.

The mixture of l.m.w. renin and extracts of the isolated tubular segments with sodium tetrathionate were similarly applied to G3000SW. As shown in Fig. 5(b), two peaks of the renin activity were observed on the chromatogram. The retention time of the renin activity was 32 min (mol. wt 60 000) and 36.5 min (mol. wt 38 000), respectively. These results suggested that renin binding substance was present in the tubular cells but not in the juxtaglomerular cells. Therefore, renin binding substance would not contribute to the biosynthesis of renin might play a role in tubular function.

Application of [¹²⁵I] renin to G3000SW. G3000SW was also used in the study of the metabolism of submaxillary renin in the mouse (Iwao *et al.*, 1982b, c, 1983a, b). Renin A was labeled with ¹²⁵I by the chloramine-T method (Michelakis *et al.*, 1974b; Iwao *et al.*, 1980). The labeled renin was separated by gel filtration on Sephadex G-25. Subsequently [¹²⁵I] renin fraction was applied to a column of G3000SW (600 × 7.5 mm i.d.). As shown in Fig. 6(a), two peaks of the radioactivity were eluted from a G3000SW column, the earlier peak of which was [¹²⁵I] renin, the later peak free ¹²⁵I. [¹²⁵I] Renin was re-chromatographed on G3000SW. The elution pattern (Fig. 6b) showed a single peak with a retention time of 21 min corresponding to a molecular weight of 38 000. More than 95% of [¹²⁵I] renin purified in such a way was precipitated by an antiserum of the mouse submaxillary renin. When a radioactive monitor (Packard, TRI-CARB RAM 7500) was connected to the HPLC system, a single radioactive peak was also

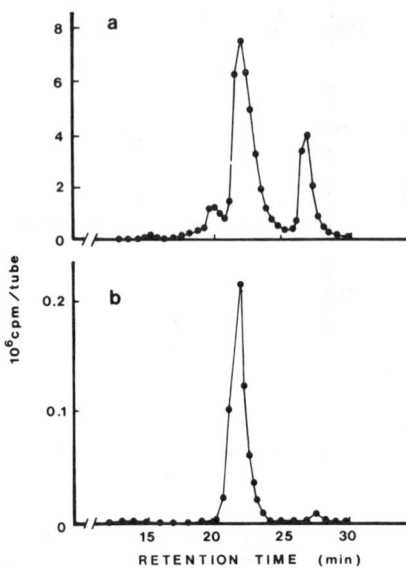

Figure 6. (a) The chromatographic pattern of [125 I] renin separated by a Sephadex G25 column on a G3000SW column. (b) The re-chromatographic profile of [125 I] renin eluted in (a) on a G3000SW column. 0.5 ml of each fraction was collected. The system consisted of a single G3000SW column (600 × 7.5 mm) and a GSWP column as a pre-column (75 × 7.5 mm), different from the double G3000SW column system used in Figs 1–5. [125 I] Renin was eluted at a retention time of 21 min.

obtained by an application of [125 I] renin on G3000SW column. Using this system, the speed and efficacy were increased for determination of the purity of radioactive renin. According to this step, labeled materials could be easily detected. These procedures used for purification of labeled materials, led to high quality and purity.

The 125 I-labeled renin was intravenously administered to male ICR mice, in a dose of 20 ng (2 μCi)/40 g body weight. At a specified time, (5, 30, 60, 180 min after injection) whole body autoradiography was performed. In other experiments, urine was collected by bladder puncture, blood was withdrawn from the inferior vena cava, and liver and kidneys were excised. Urine, plasma, supernatants of liver and kidney homogenates each at 5, 30, 60, 180 min after injection of the [125 I] renin were directly applied on a G3000SW column to separate the radioactive substance (Iwao *et al.*, 1982b, c). Figure 7(a) shows HPLC profile of plasma obtained at 5 min after injection of [125 I] renin. The radioactive peak with a retention time of 21 min indicated [125 I] renin, because the retention time of [125 I] renin was the same. This radioactivity of [125 I] renin on the chromatogram of G3000SW rapidly decreased in height after the injection, and the HPLC profile of plasma obtained at 180 min no longer showed the radioactive peak of [125 I] renin. These results suggested that intravenously administered [125] renin rather rapidly disappeared from the plasma. HPLC profiles of urine showed a very high peak of free 125 I and a slight peak of [125 I] renin (Fig. 7b). In the kidney extract, the

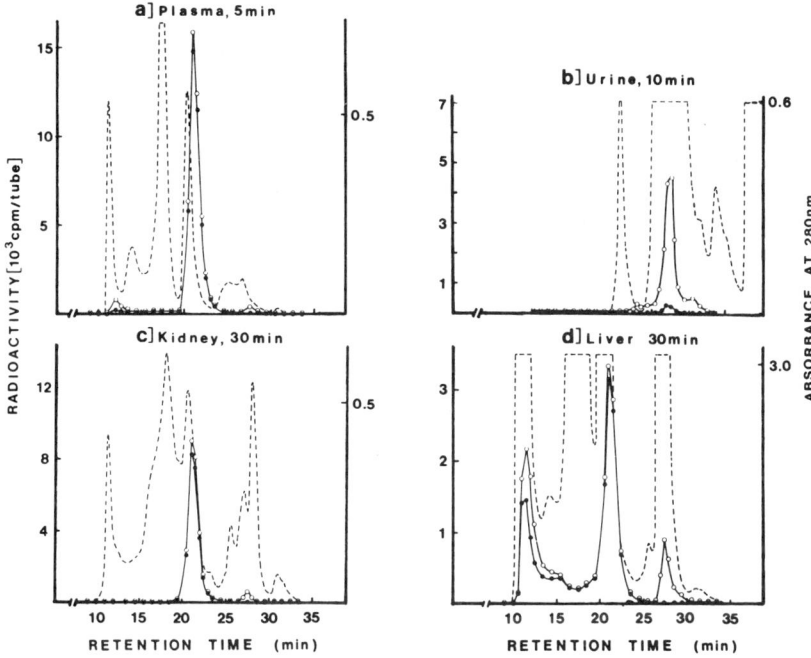

Figure 7. HPLC profile of plasma at 5 min after injection of (a) [^{125}I] renin, (b) urine at 10 min, (c) kidney at 30 min, and (d) liver at 30 min. The dotted line indicates absorbance at 280 nm, and the solid line and open circles indicate total radioactivity. The solid line and solid circles indicate the radioactivity of immunoreactive [^{125}I] renin. [^{125}I] Renin was eluted at a retention time of 21 min.

radioactivity of [^{125}I] renin eluted by HPLC on G3000SW was the highest in the kidney excised at 30 min (Fig. 7c), after which the peak of [^{125}I] renin gradually decreased. However, the radioactivity of kidney extract at 180 min after injection was still 10 times higher than that in the plasma. In the liver, the radioactivity of [^{125}I] renin on HPLC at 30 min was very small compared with plasma and kidney (Fig. 7d) and the peak decreased with time, while free ^{125}I increased with time. These results indicated that exogenously administered [^{125}I] renin in mouse was distributed from plasma predominantly into the kidney, not the liver, and coincided with the finding of light and electron microscope autoradiography that the accumulation of [^{125}I] renin in the upper area of the proximal convoluted tubule was evident, but in the liver was minimal (Iwao *et al.*, 1982b). Thus, it was demonstrated that circulating renin is re-absorbed in the proximal convoluted tubule after filtration in the glomerular capillaries. In addition, Iwao *et al.* (1983a) determined subcellular localization of exogenously administered [^{125}I] renin in mouse kidney using HPLC on G3000SW. Subcellular fractions were studied by two techniques including differential centrifugation (Morimoto *et al.*, 1981) and discontinuous sucrose density gradient centrifugation (Morimoto *et al.*, 1972). With differential centrifugation, about a half of the radioactivity of [^{125}I] renin was recovered in the heavy mitochondrial fraction and which contained the highest activity of acid phosphatase, the reference enzyme for lysosomes. Figure 8(a) shows the HPLC profile of this heavy mitochondrial fraction on G3000SW. Radioactivity showed a major peak with a retention time of 21 min, such being the same position in the original [^{125}I] renin. By discontinuous sucrose density gradient centrifugation, two peaks of radioactivity were observed in the 1.2 and 1.5 M sucrose fraction. Lysosomes were mainly included in 1.2 and 1.5 M sucrose fraction (Morimoto et al., 1981) and the HPLC profile of 1.2 M sucrose fraction showed a radioactive peak with the same retention time as the peak of renin, as did the heavy mitochondrial fraction (Fig. 8b). Application of 1.5 M sucrose fraction gave the same result. All these data demonstrated that exogenously administered renin filtered through the glomerular capillaries is re-absorbed in the proximal

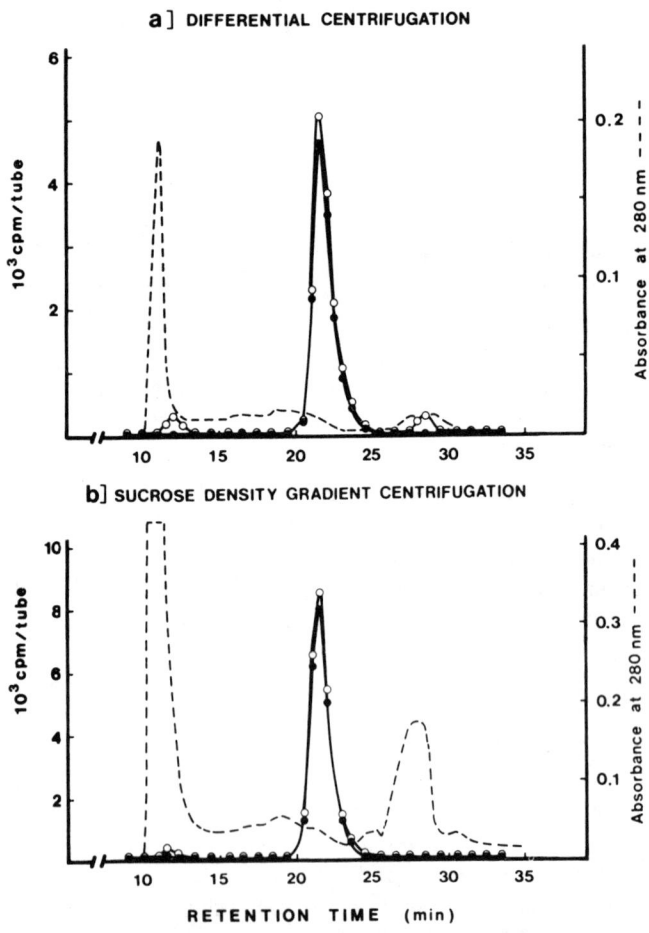

Figure 8. HPLC profile of heavy mitochondrial fraction (a) and 1.2 M sucrose fraction (b). As in Fig. 7, the solid line and solid circles indicate the radioactivity of immunoreactive [^{125}I]-renin, and the solid line and open circles total radioactivity. The broken line indicates the absorbance at 280 nm. In both Figs 8(a) and (b), a single radioactive peak with a retention time of 21 min was of [^{125}I] renin.

convoluted tubule and mainly locates in the lysosomal granules. HPLC on G3000SW proved to be a most useful technique for study of high molecular weight renin and renin binding substance, and on the metabolism of exogenous renin in mice. HPLC on G3000SW, is therefore superior to conventional chromatography with regard to separation, reproducibility, and execution time.

High-performance ion-exchange chromatography

Purification of submaxillary renin by a conventional method. Table 2 shows the result of purification of submaxillary renin by the five steps of conventional chromatography (Cohen *et al.*, 1972). The sample of renin activity eluted by chromatography on DE-cellulose (step 4) was separated into four enzymatically active fractions by the final chromatography on CM-cellulose. These fractions were termed renin A, B, C, and D. Renin A was of the major peak and renin C was of the second. Each renin A and C showed a single band with discrete migrations on disc electrophoresis on polyacrylamide gel without SDS (Fig. 9). In the B fraction, one of the two bands showed the same migration as renin A, thereby indicating that renin A contaminated the renin B fraction. As shown in Fig. 10, the electrophoresis of these homogenous renin A and C on polyacrylamide gel with SDS resulted in two protein bands with the same migration and which consisted of heavy chain with molecular weight of 34 000 and light chain (Misono and Inagami, 1982; Corvol *et al.*, 1983a). The specific activity of renin A and C was 25 mg angiotensin I/h mg protein and 18 mg angiotensin I/h mg protein, respectively.

Table 2. Summary of purification of renin from mouse submaxillary gland

Preparation	Total protein (mg)	Total activity (mgAI/h)[a]	Specific activity[b]	Yield (%)	Purification
Ammonium sulfate	832	1736	2.09	100	1
Sephadex G-100	159	974	6.13	56	2.9
DEAE-cellulose	121	783	6.47	45	3.1
CM-cellulose	30	479	15.9	28	7.6
DEAE-cellulose	12	245	20.4	14	9.8
Final CM-cellulose					
Renin A	4.5	114	25.3	6.5	12.1

[a] mg Angiotensin I/h.
[b] mg AI/h/mg protein.

Comparison between IEX535CM and CM-cellulose. The initial buffer (buffer A) used was 0.05 M sodium acetate, pH 5.4. The gradient buffer (buffer B) was 0.05 M sodium acetate containing 1 M sodium acetate, pH 6.9. A linear gradient of sodium acetate was given by increasing 1% of buffer B per minute. All elution conditions are briefly mentioned in Fig. 11. When the second DE-cellulose treated fraction (step 4) was chromatographed on IEX535CM (150 X 6 mm) instead of CM-cellulose,

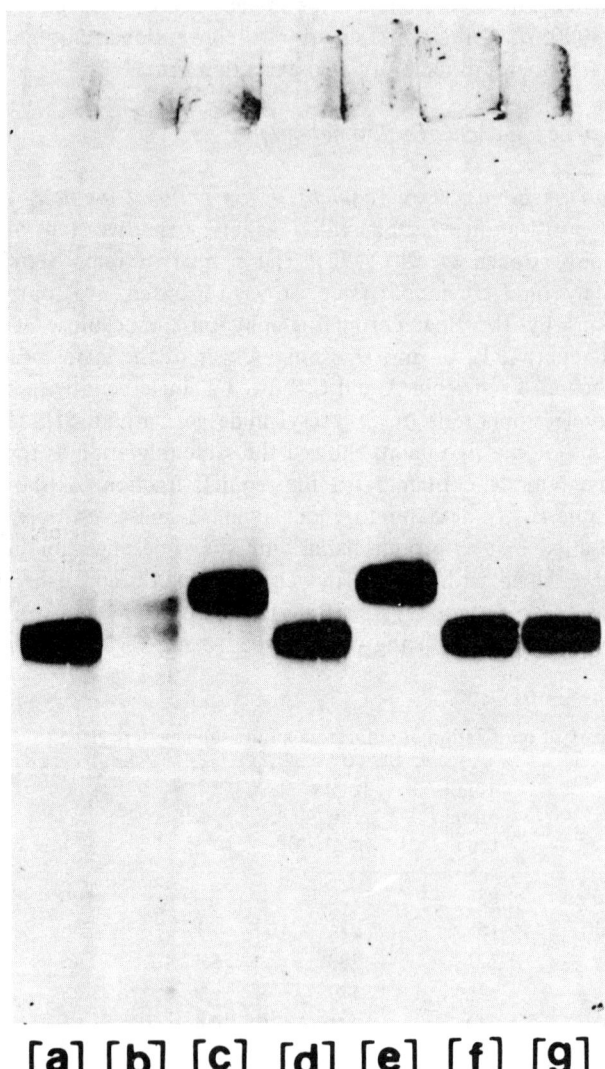

[a] [b] [c] [d] [e] [f] [g]

Figure 9. The disc electrophoresis of isoenzymes of renin on 7% polyacrylamide gel without sodium dodecyl sulfate. (f) and (g) showed a single band of renin A_1 and renin A_2, respectively. (a) Renin A, (b) renin B, and (c) renin C were purified by conventional methods. On the other band, renin A (d) and renin C (e) were purified by a novel two-step method. All renin A_1 (f), renin A_2 (g), renin A (a), and renin A (d) gave a single band with the same value of R_f. Both renin C (c) and renin C (e) also gave a single band with the same value of R_f. Renin B (b) had two different bands, one of the two bands had the same R_f as renin A, which indicated that renin B was contaminated with renin A.

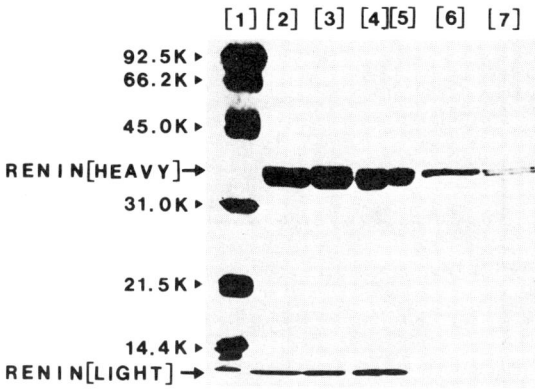

Figure 10. The slab electrophoresis of isoenzymes of renin on 12% polyacrylamide gel with sodium dodecyl sulfate in the presence of 2-mercaptoethanol. Each number means (1) standard proteins, (2) renin A and (3) renin C purified by conventional methods, (4) renin A and (5) renin C purified by a novel method, (6) renin A_1 and (7) renin A_2. Upper arrow indicates heavy chain of renin, and the lower arrow the light chain.

it was separated into three protein peaks (peaks 1, 2, and 3), the retention time of which is 27, 32, and 41 min, respectively (Fig. 11a). All peak fractions had renin activity. To compare the similarity among renin A, C, D and the three fractions (peaks 1, 2, 3) eluted on IEX535CM, Renin A, C, D was individually applied to a IEX535CM column. In each HPIEC application, a single peak was observed, thus supporting the idea that they were all pure. The retention time of renin A is around 27 min, identical to that of the major peak (peak 1) (Fig. 11b). Renin C and D were eluted in the same retention time as peak 2 and 3, 32 and 41 min, respective 'v (Fig. 11c and d). The chromatographs of each renin preparation were perforr ِd several times, and their retention times were always the same, under the same conditions of elution. The recovery of renin A (0.4 mg) from IEX535CM was about 60% (Table 3). Thus, by HPIEC on IEX535CM, the separation into iso-enzymes of submaxillary gland renin could be performed with high recovery within only 1 h, while the conventional chromatography on CM-cellulose required one day for the separation. Excellent reproducibility was always obtained.

HPIEC of submaxillary gland renin on SP-5PW. The chromatography of sub-maxillary gland renin on SP-5PW (75 × 75 mm i.d.), another cation exchanger which is not silica-based, but polymer-based and chemically stable was also carried out. When 20 mM phosphate buffer, pH 7.5, was used, renin A was not adsorbed to the column. In the case of 20 mM phosphate buffer, pH 6.0, renin A was adsorbed and subsequently its elution profile with a 0–0.5 M NaCl gradient in 20 mM phosphate buffer (pH 6.0) showed an unsymmetric protein peak with tailing and a broad width. Although we used a linear gradient of 0–1 M NaCl in 50 mM sodium acetate buffer, pH 5.5, renin A, B, C, and D were not completely separated. These results suggest that the chromatography of renin on SP-5PW is not so useful as on IEX535CM.

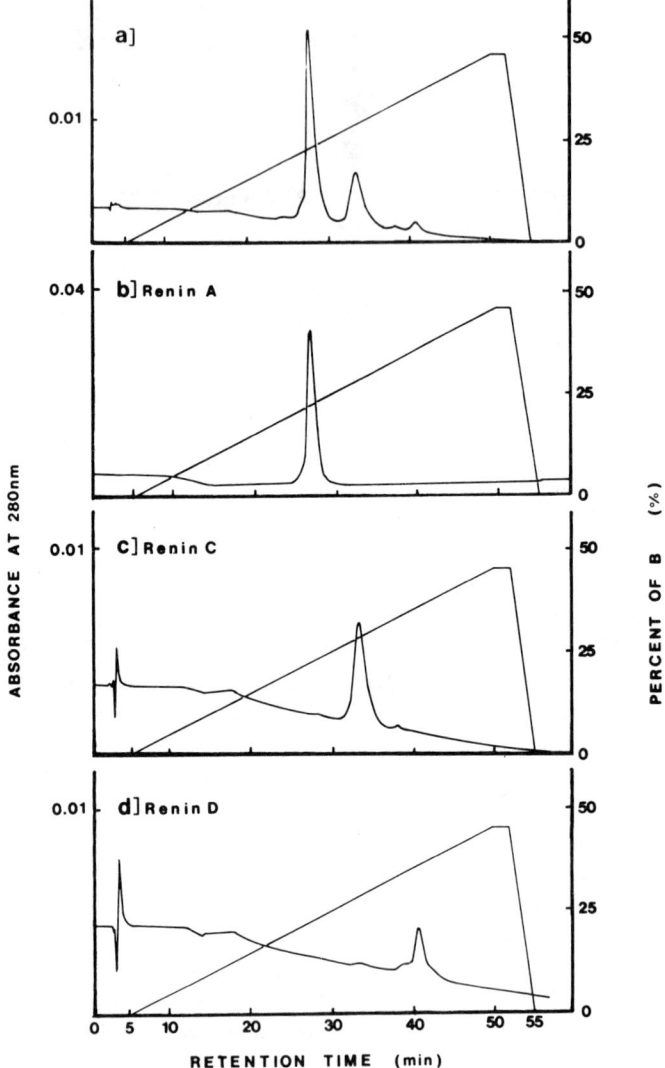

Figure 11. HPLC profile of the second DE-cellulose-treated renin activity (a), renin A (b), renin C (c), and renin D (d) on a IEX535CM. A initial buffer (buffer A) was 0.05 M sodium acetate, pH 5.4 and a gradient buffer (buffer B) was 0.05 M sodium acetate, pH 6.9, containing 1 M sodium acetate. Each sample was applied to an IEX535CM column (150 × 6 mm) equilibrated with 0.05 M sodium acetate buffer, pH 5.4 and then eluted with a linear gradient from buffer A to buffer B the flow rate was 1.0 ml/min. Protein was detected by the absorbance at 280 nm. All the chromatographic runs were performed at room temperature of about 25°C. The chart speed was 30 cm/min.

HPIEC of submaxillary gland renin on DEAE-3SW. Renin A, C and D were eluted at the same retention time by the chromatography on DEAE-3SW (7.5 × 75 mm i.d.), with a linear gradient of NaCl in 0.02 M Tris-HCl buffer (pH 7.4). Even with a low gradient slope, their retention times were equal, and isoenzymes

Table 3. Recovery of renin activity

Column	Recovery (%)
IEX-535CM	60
DEAE-3SW	60
DEAE-5PW	62

Renin A of 0.4 mg was applied to each column. Renin activity was determined by radioimmunoassay of angiotensin I, as described in the text.

of the submaxillary gland renin could not be separated. Figure 12(a and b) shows the elution profile of renin A and C on DEAE-3SW, the retention time being about 24 min.

However, when 0.02 M Tris-acetate buffer pH 7.4 containing 1 M sodium acetate was used as a gradient buffer, a different result was obtained. Renin A was separated into two protein peaks, both of which had renin activity and were designated as renin A_1, and A_2. The retention times were 36 and 38 min, respectively (Fig. 13).

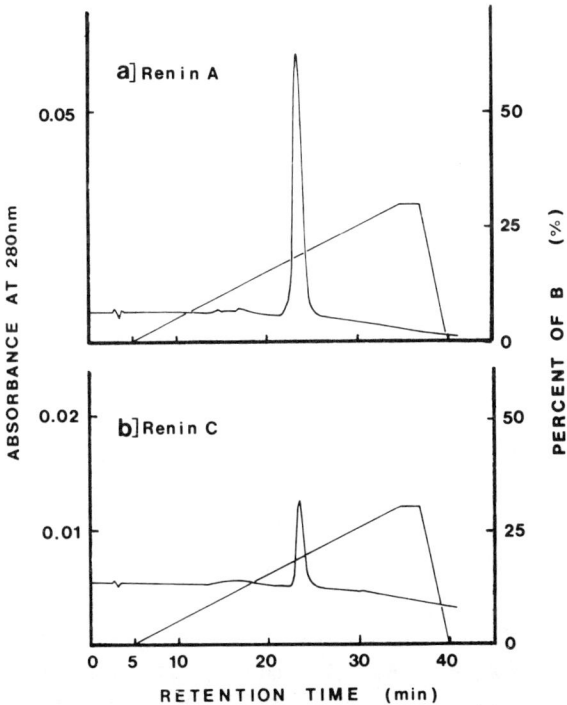

Figure 12. HPLC profile of renin A (a) and renin C (b) on DEAE 3 SW (75 × 7.5 mm). An initial buffer (buffer A) was 0.02 M Tris-HCl, pH 7.4 and a gradient buffer (buffer B) 0.02 M Tris-HCl, pH 7.4 containing 1 M NaCl. HPLC was performed with a linear gradient from buffer A to buffer B at a constant flow rate of 1.0 ml/min.

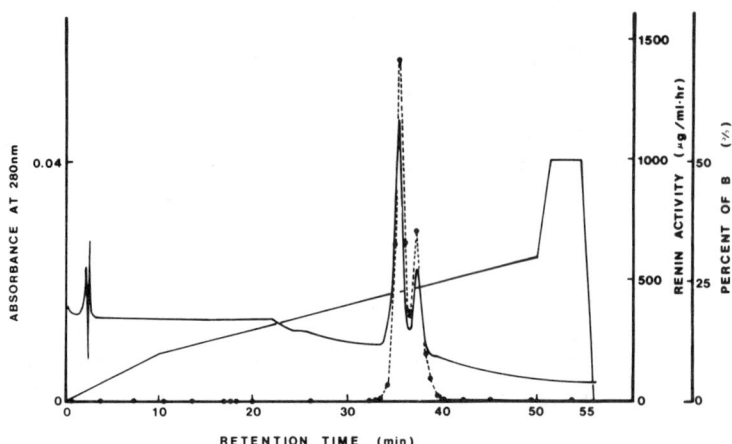

Figure 13. HPLC profile of renin A on DEAE-3 SW (75 × 7.5 mm). The gradient elution was performed with a linear gradient from 0.02 M Tris-acetate, pH 7.4 (buffer A) to 0.02 M Tris-acetate (pH 7.4) containing 1 M sodium acetate (buffer B). The solid line indicates the absorbance at 280 nm, and the broken line and solid circles indicate renin activity. The chromatographic runs were carried out at a constant flow rate of 1.0 ml/min, at room temperature of about 25°C, and the protein eluted was detected at 280 nm.

The specific activities of renin A_1 and renin A_2 were 25 mg angiotensin I/h mg protein and 20 mg angiotensin I/h mg protein, respectively. The disc electrophoresis of renin A_1 and A_2 on polyacrylamide gel without SDS demonstrated that each gave a single band and with the same migration (Fig. 9). When renin A_1 and A_2 were examined by SDS-polyacrylamide gel electrophoresis, both renin A_1 and A_2 showed two protein bands with the same mobility, namely heavy and light chains, but the light chain was hard to identify because of small amounts of applied samples (Fig. 10). Thus, renin A, which was electrophoretically homogenous and eluted as a single peak by chromatography on IEX535CM, could be separated into two components by chromatography on DEAE-3SW with a linear gradient of sodium acetate, but not NaCl. These results indicate that the choice of buffer is a very important factor for the separation by HPIEC and that this separation is sometimes superior to that using disc electrophoresis on polyacrylamide gel.

HPIEC of submaxillary gland renin on DEAE-5PW. DEAE-5PW (7.5 × 75 mm i.d.), another DEAE column with a polymer-base, is chemically more stable than DEAE-3SW and so can be used in a wide pH range (pH 2–10). The pore size of the former (1000 Å) is larger than that of the latter (250 Å). The chromatography of renin A on DEAE-5PW was performed to examine whether renin A could be separated into two components by chromatography on DEAE-5PW as on DEAE-3SW. The chromatography of renin A on DEAE-5PW was performed with the same two kinds of gradient buffers as on DEAE-3SW. In the case of elution condition consisting of 0.02 M Tris-HCl, pH 7.4 and 0.02 M Tris-HCl, pH 7.4 containing 1 M NaCl, renin A with a shoulder was eluted with a linear gradient of NaCl (Fig. 14a).

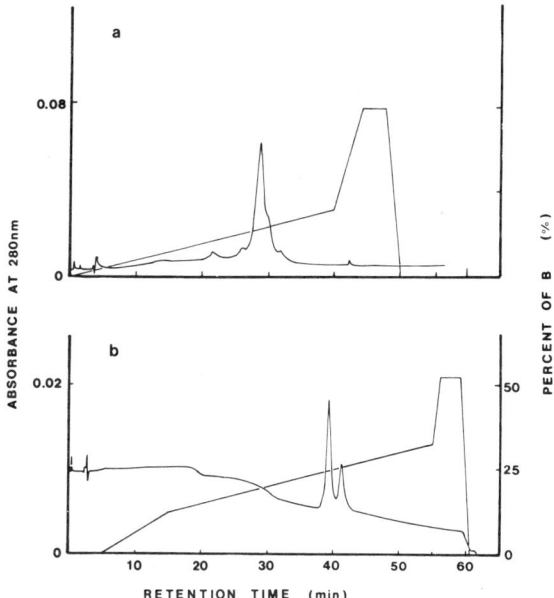

Figure 14. HPLC profile of renin A on DEAE-5PW (75 × 7.5 mm) with two kinds of linear gradient, namely with a linear gradient of NaCl in 0.02 M Tris-HCl (a) and with a linear gradient of sodium acetate in 0.02 M Tris-acetate (b). (a) Shows the chromatogram of renin A with a shoulder eluted with a gradient of NaCl. (b) Shows the complete separation into renin A_1 and renin A_2 with a gradient of sodium acetate, and each retention time of renin A_1 and renin A_2 was a little later than that in the case of DEAE 3SW as the gradient elution was started 5 min after the application of the sample.

The elution pattern of renin A on DEAE-5PW with a linear gradient of sodium acetate was consistent with that on DEAE-3SW. The chromatography of renin A on DEAE-5PW with a gradient of NaCl resulted in the elution pattern with a shoulder. This was not observed on DEAE-3SW, and the separation into two enzymatically active components with a linear gradient of sodium acetate on DEAE-5PW was the same result as that on DEAE-3SW (Fig. 14b). The chromatography of renin A on DEAE-5PW showed a better separation with a linear gradient of sodium acetate than of NaCl. The recovery of renin from DEAE-5PW is equal to that from DEAE-3SW (60%), as shown in Table 3. Thus, DEAE-5PW has at least the same separating ability and high recovery as DEAE-3SW and so can be used for the chromatography of renin because of its chemical ability.

Purification of mouse submaxillary gland renin by the two-step procedure
HPLC separations have the advantages of high speed (within an hour), excellent recovery, capability of automation, and do not require subambient elution conditions, as the elution time is much shorter compared to low-pressure column chromatography. Thus, HPLC techniques are useful not only in the fractionation of proteins but also for protein purification.

As mentioned in Materials and Methods, the homogenate of mouse submaxillary gland was applied to a column of pepstatin-aminohexyl sepharose, and the concentrated renin from the column was applied to a IEX535CM column. Figure 15(a) illustrates the elution pattern of 0.1 mg of this sample on IEX535CM with a linear gradient of NaCl in 0.05 M sodium acetate buffer. Figure 15(b) shows the elution profile in the case of increasing sample loads from 0.1 to 3.0 mg. The patterns are similar, thereby indicating that 3.0 mg of sample loading does not exceed the limitation of the IEX535CM column. The major fraction was eluted at 27 min, identical to renin A, and the second major fraction at 32 min was consistent with renin C. On polyacrylamide gel electrophoresis without SDS, each major fraction and the second major fraction gave a single band which migrated with the same distance as renin A and C, respectively (Fig. 9). Electrophoresis of these fractions on polyacrylamide gel with SDS also gave the same result as renin A and C (Fig. 10). These data supported the idea that the major fraction is renin A and the second major fraction renin C. A summary of the purification using this method is listed in Table 4. As a result, from 7 g of mouse submaxillary gland, 4.5 mg of renin A and 1.4 mg of renin C were obtained. The recovery of renin A by two step purification was approximately twice that by the method of Cohen et al. (1972) (Tables 2 and 4). Thus, submaxillary renin was successfully purified by these two steps of chromatography, in a shorter time and with a higher recovery than by the method of Cohen et al.

The application of this renin A on either DEAE-3SW or DEAE-5PW with a linear gradient of sodium acetate in 0.02 M Tris-acetate, pH 7.4 gave the same result of separation into renin A_1 and A_2 as that of renin A purified by the conventional method.

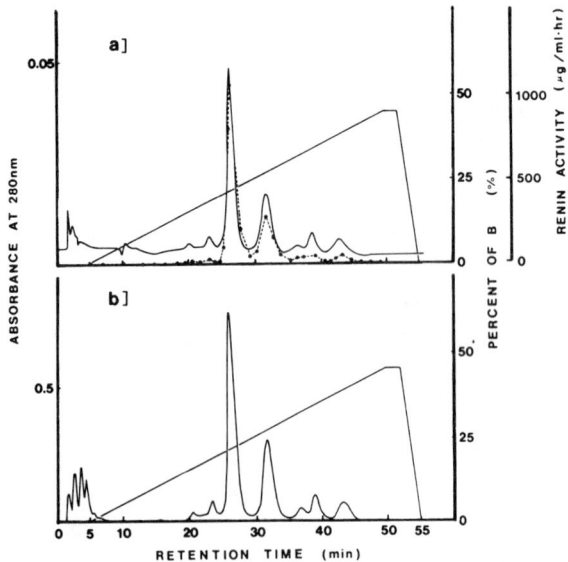

Figure 15. HPLC profile of the concentrated material of renin activity from pepstatin affinity column on IEX535CM (150 × 6 mm). The sample loading was 0.1 (a) and 3 mg (b). The solid line indicates absorbance at 280 nm and the broken line and solid circles renin activity.

Table 4. Purification of renin from mouse submaxillary gland using two steps of chromatography

Preparation	Total protein (mg)	Total activity (mg AI/h)	Specific activity (mg AI/h mg protein)	Yield (%)	Purification (*n*-fold)
Crude extract	407	957	2.4	100	1
Affinity column IEX535CM	14.8	342	23.1	35.7	9.8
Renin A	4.5	112	25.0	12	10.7

However, the success is attributed mostly to a pepstatin-aminohexyl-sepharose column which is currently the most powerful tool for the purification of renin. Combining affinity chromatography like pepstatin and specific antibody with high-performance ion-exchange chromatography will probably be a most potent technique for the purification of renin, not only from mouse submaxillary gland but also from the kidney, an organ in which the renin content is extremely low.

High-performance gel permeation and ion-exchange chromatography excel in speed, separation, reproducibility, capability of automation, and in most cases need not be done in a cold room, as compared with conventional chromatographic methods. In addition the results are sometimes superior to these obtained with electrophoresis. Drawbacks are the limitation of sample loading, chemical instability (in the case of silica-supported column), and the cost. When these factors are given attention, high-performance liquid chromatography should be a most powerful tool.

CONCLUSIONS

This chapter presents studies on high-performance liquid chromatography of renin in gel permeation on G3000SW columns and in ion exchange including IEX535CM, SP-5PW, DEAE-3SW, and DEAE-5PW columns. All these columns were obtained from Toyo Soda (Yamaguchi, Japan).

High-performance gel permeation chromatography on G3000SW has advantages of excellent separation, high speed, and the high reproducibility and was useful for studies of high molecular weight renin and renin binding protein and metabolism of renin.

Mouse submaxillary gland renin was purified by the method of Cohen *et al.* (1972) and was separated into renin A, B, C, and D by the final chromatographic step on CM-cellulose. When renins A, C, and D were individually applied to a IEX535CM column, they were eluted with a linear gradient of NaCl in 0.05 M sodium acetate buffer at a constant elution time of 27, 31, and 41 min, respectively. The reproducibility was excellent. The application of renin on SP-5PW, another cation exchanger was unsuccessful. Renin A, which gave a single band on polyacrylamide gel electrophoresis was separated into two components on the chromatography of either DEAE-3SW or DEAE-5PW with a linear gradient of sodium acetate in 0.02 M Tris-acetate, pH 7.4. These two components were enzymatically active and gave a single band with the same migration distance on polyacrylamide gel electrophoresis.

Mouse submaxillary gland renin could be purified by another method using IEX535CM. It was a two-step procedure including pepstatin affinity chromatography and then HPLC on IEX535CM. On the chromatogram of IEX535CM, several peaks of the renin activity were seen. The major protein peak was eluted at the same retention time as renin A and the second major peak at the same retention time as renin C. Thus, by these two steps of chromatography, renin A and renin C were also purified, and this method proved to be superior to the method of Cohen *et al.* (1972) with regard to recovery and the time required.

Acknowledgement
We thank M. Ohara for helping us prepare this article.

REFERENCES

Aoyagi, T., Morishima H., Nishizawa R., Kunimoto, S., Takeuchi, T., and Umezawa H. (1972). Biological activity of pepstatins, pepstanone A and partial peptides on pepsin, cathepsin D and renin. *J. Antibiot. 25*, 689–694.

Bensadoun, A. and Weinstein, D. (1976). Assay of proteins in the presence of interfering materials. *Anal. Biochem. 70*, 241–250.

Bing, J. and Poulsen, K. (1971). The renin system in mice. *Acta. Pathol. Microbiol. Scand. (A) 79*, 134–138.

Bing, J., Poulsen, K., Hackenthal, E., Rix, E., and Taugner, R. (1980). Renin in the submaxillary gland: a review. *J. Histochem. Cytochem. 28*, 874–880.

Boyd, G. W. (1974). A protein-bound form of porcine renal renin. *Cir. Res. 35*, 426–438.

Braun-Menendez, E., Fasciolo, J. C., Leloir, L. F., and Munoz, J. M. (1940). The substance causing renal hypertension. *J. Physiol. Lond. 98*, 283–298.

Campbell, W. B., Brooks, S. N., and Pettinger, W. A. (1974). Angiotensin II- and angiotensin III-induced aldosterone release *in vivo* in the rat. *Science 184*, 994–996.

Catanzaro, D. F., Mullins, J. J., and Morris, B. J. (1983). The biosynthetic pathway of renin in mouse submandibular gland. *J. Biol. Chem. 258*, 7364–7368.

Chang, S. H., Gooding, K. M., and Regnier, F. E. (1976). High-performance liquid chromatography of proteins. *J. Chromatogr. 125*, 103–114.

Cohen, S., Taylor, J. M., Murakami, K., Michelakis, A. M., and Inagami, T. (1972). Isolation and characterization of renin like enzymes from mouse submaxillary glands. *Biochemistry 11*, 4286–4293.

Corvol, P., Devaux, C., and Menard, J. (1973). Pepstatin, an inhibitor for renin purification by affinity chromatography. *FEBS Lett. 34*, 189–192.

Corvol, P., Devaux, C., Ito, T., Sicard, P., Ducloux, J., and Menard, J. (1977). Large scale purification of hog renin: physicochemical characterization. *Circ. Res. 41*, 616–622.

Corvol, P., Panthier, J. J., Foote, S., and Rougeon, F. (1983a). Structure of the mouse submaxillary gland renin precursor and a model for renin processing. Arthur C. Carcoran Memorial Lecture. *Hypertension 5 (suppl. I)*, 3–9.

Corvol, P., Panthier, J. J., Souhrier, F., Menard, J., Sicard, P., and Rougeon, F. (1983b). Mouse submaxillary renin: a useful model for the study of renal renin. *J. Hypert. 1 (suppl. 1)*, 3–7.

Davis, B. J. (1964). Disc electrophoresis. *Ann. NY Acad. Sci. 121*, 404–427.

Day, R. P. and Leutscher, J. A. (1974). Big renin: a possible prohormone in kidney and plasma of a patient with Wilm's tumor. *J. Clin. Endo. Meta. 40*, 923–926.

Day, R. P., Leutscher, J. A., and Gonzales, C. M. (1975). Occurrence of big renin in human plasma, amniotic fluid and kidney extracts. *J. Clin. Endo. Meta. 40*, 1078–1084.

Devaux, C., Menard, J., Sicard, P., and Corvol, P. (1976). Partial characterization of hog renin purified by affinity chromatography. *Eur. J. Biochem. 64*, 621–627.

Dzau, V. J., Slater, E. E., and Haber, E. (1979). Complete purification of dog renal renin. *Biochemistry 18*, 5224–5228.

Edelman, R. and Hartroft, P. M. (1961). Localization of renin in juxtaglomerular cells of rabbit and dog through the use of the fluorescent-antibody technique. *Circ. Res. 9*, 1069–1077.

Freedlender, A. E., Fyhrquist, F., and Hollemans, H. J. G. (1974). Renin and the angiotensins. In: *Methods of Hormone Radioimmunoassay*, B. M. Jaffe and H. R. Behrman (Eds). Academic Press, New York, pp. 455–469.

Fukano, K., Komiya, K., Sasaki, H., and Hashimoto, T. (1978). Evaluation of new supports for high pressure aqueous gel permeation chromatography: TSK–GEL SW type column. *J. Chromatogr 166*, 47–54.

Funakawa, S., Funae, Y., and Yamamoto, K. (1978). Conversion between renin and high-molecular-weight renin in the dog. *Biochem. J. 176*, 977–981.

Galen, F. X., Devaux, C., Guyenne, T., Menard, J., and Corvol, P. (1979). Multiple forms of human renin: purification and characterization. *J. Biol. Chem. 254*, 4848–4855.

Haber, E., Koerner, T., Page, L. B., Kliman, B., and Purnode, A. (1969). Application of a radioimmunoassay for angiotensin I to the physiological measurement of plasma renin activity in normal human subjects. *J. Clin. Endocr. 29*, 1349–1355.

Haber, E. and Carlson, W. (1982). The biochemistry of the renin angiotensin system. In: *Hypertension; Physiopathology and Treatment*, J. Genest, O. Kuchel, P. Hamet, and M. Cantin (Eds). McGraw-Hill, pp. 171–184.

Hass, E., Lamfrom, H., and Goldblatt, H. (1953). Isolation and purification of hog renin. *Arch. Biochem. Biophys. 42*, 368–380.

Higaki, J., Hirose, S., Ogihara, T., Imai, N., Kisaragi, M., Murakami, K., and Kumahara, Y. (1982). A novel purification method of human renin. *Life Sci. 32*, 1591–1591–1598.

Ho, S., Izumi, H., and Michelakis, A. M. (1982). The purification and characterization of multiple forms of mouse submaxillary gland renin. *Biochim. Biophys. Acta 717*, 405–413.

Ikemoto, F., Takaori, K., Iwao, H., and Yamamoto, K. (1982). Intrarenal localization of renin binding substance in rats. *Life Sci. 31*, 1011–1016.

Imbelt-Teboul, M., Chabardes, D., Montegut, M., Clique, A., and Morel, F. (1978). Vasopressin-dependent adenylate cyclase activities in the rat kidney medulla: evidence for two separate sites of action. *Endocrinology 102*, 1254–1261.

Inagami, T. and Murakami, K. (1977). Pure renin: isolation from hog kidney and characterization. *J. Biol. Chem. 252*, 2978–2983.

Inagami, T., Hirose, S., Murakami, K., and Matoba, T. (1977). Native form of renin in the kidney. *J. Biol. Chem. 252*, 7733–7737.

Iwao, H., Lin, C.-S., and Michelakis, A. M. (1980). Effect of adrenergic agonists on big and small renin. *Am. J. Physiol. 238*, E416–E420.

Iwao, H., Minami, T., Ikemoto, F., Takaori, K., Nakamura, N., and Yamamoto, K. (1982a). High molecular weight renin identified in cytosol fractions of mouse renal cortices. *Biochem. Biophys. Res. Commun. 106*, 933–939.

Iwao, H., Nakamura, N., Ikemoto, F., Yamamoto, K., Mizuhira, V., Ono, M., and Sugiura, Y. (1982b) Distribution of exogenously administered renin in mouse kidney. *Clin. Exp. Hypertension A4*, 2449–2456.

Iwao, H., Nakamura, N., Takaori, K. Ikemoto, F., and Yamamoto, K. (1982c). Metabolism of ^{125}I-labeled renin. *Jpn. J. Pharmacol. 32* (suppl.), 60.

Iwao, H., Nakamura, N., Ikemoto, F., and Yamamoto, K. (1983a). Subcellular localization of exogenously administered renin in mouse kidney. *Jpn. Cir. J. 47*, 1198–1202.

Iwao, H., Nakamura, N., Ikemoto, F., and Yamamoto, K. (1983b). Whole body autoradiographic distribution of exogenously administered renin in mice. *J. Histochem. Cytochem. 31*, 776–782.

Kato, Y., Komiya, K., Sasaki, H., and Hashimoto, T. (1980a). High-speed gel filtration of proteins in 6 M guanidine hydrochloride on TSK-GEL SW columns. *J. Chromatogr. 193*, 458–463.

Kato, Y., Komiya, K., Sawada, Y., Sasaki, H., and Hashimoto, T. (1980b). Purification of enzymes by high-speed gel filtration on TSK-GEL SW columns. *J. Chromatogr. 190*, 305–310.

Kawamura, M., Ikemoto, F., Funakawa, S., and Yamamoto, K. (1979). Characteristics of a renin-binding substance for the conversion of renin into a higher-molecular-weight form in the dog. *Clin. Sci. 57*, 345–350.

Laemmli, U. K. (1970). Cleavage of structural proteins during the assembly of the head of bacteriophage T4. *Nature 227*, 680–685.

Leckie, B. J. and McConnel, A. (1975). A renin inhibitor from rabbit kidney: conversion of a large inactive renin to a smaller active enzyme. *Cir. Res. 36*, 513–519.

Lowry, O. H., Rosebrough, N. J., Farr, A. L., and Randall, R. J. (1951). Protein measurement with the Folin phenol reagent. *J. Biol. Chem. 193*, 265–275.

Matoba, T., Murakami, K., and Inagami, T. (1978). Rat renin: purification and characterization. *Biochim. Biophys. Acta 526*, 560–571.

Menard, J. and Catt, K. (1972). Measurement of renin activity, concentration and substrate in rat plasma by radioimmunoassay of angiotensin I. *Endocrinology 90*, 422–430.

Michelakis, A. M., Cohen, S., Taylor, J., Murakami, K., and Inagami, T. (1974a). Studies on the characterization of pure submaxillary gland renin. *Proc. Soc. Exp. Biol. Med. 147*, 118–121.

Michelakis, A. M., Yoshida, H., Menzie, J., Murakami, K., and Inagami, T. (1974b). A radioimmunoassay for the direct measurement of renin in mice and its application to submaxillary gland and kidney studies. *Endocrinology 94*, 1101–1105.

Misono, K. S. and Inagami, T. (1980). Characterization of the active site of mouse submaxillary gland renin. *Biochemistry 19*, 2616–2622.

Misono, K. S., Holladay, L. A., Murakami, K., Kuromizu, K., and Inagami, T. (1982a). Rapid and large-scale purification and characterization of renin from mouse submaxillary gland, *Arch. Biochem. Biophys. 217*, 574–581.

Misono, K. S. and Inagami, T. (1982). Structure of mouse submaxillary gland renin: identification of two disulfide-linked polypeptide chains and the complete amino acid sequence of the light chain. *J. Biol. Chem. 257*, 7536–7540.

Misono, K. S., Chang, J. J., and Inagami, T. (1982b). Amino acid sequence of mouse submaxillary gland renin. *Biochemistry 79*, 4858–4862.

Morimoto, S., Yamamoto, K., and ueda, J. (1972). Isolation of renin granules from the dog kidney cortex. *J. Appl. Physiol. 33*, 306–311.

Morimoto, S., Nakmura, N., Fukuhara, A., Abe, R., and Matsumura, Y. (1981). Subcellular localization of renin in the mouse kidney. *Chem. Pharm. Bull. 29*, 849–854.

Murakami, K., Inagami, T., Michelakis, A. M., and Cohen, S. (1973). An affinity column for renin. *Biochem. Biophys. Res. Commun. 54*, 482–487.

Murakami, K. and Inagami, T. (1975). Isolation of pure and stable renin from hog kidney. *Biochem. Biophys. Res. Commun. 62*, 757–763.

Murakami, K., Suzuki, F., Morita, N., Ito, H., Okamoto, K., Hirose, S., and Inagami, T. (1980). High molecular weight renin in stroke-prone spontaneously hypertensive rats. *Biochim. Biophys. Acta 622*, 115–122.

Nakajima, T., Oshima, G., Yeh, H. S. J., Igie, R., and Erdos, E. G. (1973). Purification of the angiotensin I-converting enzyme of the lung. *Biochim. Biophys. Acta 315*, 430–438.

Page, I. H. and Helmer, O. M. (1940). Angiotonin-activator, renin- and angiotonin-inhibitor and the mechanism of angiotonin tachyphylaxis in normal, hypertensive and nephrectomized animals. *J. Exp. Med. 71*, 485–519.

Panthier, J. J., Foote, S., Chambraud, N., Strasberg, A. D., Corvol, P., and Rougeon, F. (1982). Complete amino acid sequence and maturation of the mouse submaxillary gland renin precursor. *Nature 298*, 90–92.

Peart, W. S., Lloyd, A. M., Thatcher, G. N., Lever, A. F., Payne, N., and Stone, N. (1966). Purification of pig renin. *Biochem. J. 99*, 708–716.

Poulsen, K., Vuust, J., Lykkegaard, S., Nielsen, A. H., and Lund, T. (1979). Renin is synthesized as a 50 000 single-chain polypeptide in cell-free translation systems. *FEBS Lett. 98*, 135–138.

Regnier, F. E. (1982). Review: high-performance ion-exchange chromatography of proteins; the current status. *Anal. Biochem. 126*, 1–7.

Regnier, F. E. and Gooding, K. M. (1980). Review: high-performance liquid chromatography of proteins. *Anal. Biochem. 103*, 1–25.

Rougeon, F., Chambraud, B., Foote, S., Panthier, J. J., Nageotte, R., and Corvol, P. (1981). Molecular cloning of a mouse submaxillary gland renin CDNA fragment. *Proc. Natl. Acad. Sci. 78*, 6367–6371.

Sagnella, G., Price, R., and Peart, W. (1980). Subcellular distribution and storage form of rat renal renin. *Hypertension 2*, 595–603.

Skeggs, L. T., Darer, F. E., Kahn, J. R., Lentz, E., and Levine, M. (1980). Biochemistry of the renin-angiotensin system and its role in hypertension. In: *Topics in Hypertension*, J. H. Laragh (Ed.). Yorke Medical Books, pp. 1–20.

Slater, E. E. and Strout, H. V. Jr (1981). Pure human renin: identification and characterization of two major molecular weight forms. *J. Biol. Chem. 256*, 8164–8171.

Stockigt, J. R., Dennis, R., and Biglieri, E. G. (1971). Determination of plasma renin concentration by angiotensin I immunoassay: diagnostic import of precise measurement of subnormal renin in hyperaldosteronism. *Cir. Res. 38 & 39 (suppl. II)*, 175–189.

Suzuki, F. and Nakamura, Y., Nagata, Y., Ohsawa, T., and Murakami, K. (1981). A rapid and large scale isolation of renin from mouse submaxillary gland by pepstatin-aminohexyl-agarose affinity chromatography. *J. Biochem. 89*, 1107–1112.

Takaori, K., Ikemoto, F., and Yamamoto, K. (1981). Biochemical properties of the renin binding substance of rat kidney. *Clin. Exp. Hypertension 3*, 991–1000.

Tobian, L., Tanecek, J., and Tomboulian, A. (1959). Correlation between granulation of juxtaglomerular cells and extractable renin in rats with experimental hypertension. *Proc. Soc. Exp. Biol. Med. 100*, 94–96.

Umezawa, H. (1973). Chemistry of enzyme inhibitors of microbial origin. *Pure. Appl. Chem. 33*, 129–144.

Varik, N. D. and Taen, E. C. Jr (1982). Review-separation and measurement of isoenzymes and other proteins by high-performance liquid chromatography. *J. Chromatogr. 228*, 1–31.

Vanecek, G. and Kegnier, F. E. (1980). Variables in the high-performance anion-exchnge chromatography of proteins. *Anal. Biochem. 109*, 345–353.

Waldhausl, W. K., Lucas, C. P., Conn, J. W., Lutz, J. H., and Cohen, E. L. (1970). Studies on the partial isolation of human renin. *Biochim. Biophys. Acta 221*, 536–548.

Yokosawa, H., Halladay, L. A., Inagami, T., Hass, E., and Murakami, K. (1980). Human renal renin-complete purification and characterization. *J. Biol. Chem. 255*, 3498–3502.

Progress in HPLC, Vol. 1, pp. 133–147
Parvez *et al.* (Eds)
© 1985 VNU Science Press

Enzyme purification by high-performance ion-exchange liquid chromatography

FREDERICK B. RUDOLPH, BRUCE F. COOPER, and JOAN GREENHUT

Department of Biochemistry, Rice University, PO Box 1892, Houston, TX 77251, USA

INTRODUCTION

Enzyme purification involves the separation of a single protein from a mixture which may contain over a thousand components. Prior to the 1950s, most protein purification procedures were based upon changing ionic strength, dielectric constant, or other solution properties. Many enzymes were not stable to such techniques or were present in such low concentrations that purification was not possible. The introduction of ion exchange and molecular sieve supports presented an opportunity to exploit size and charge differences in biological molecules to effect their purification on a preparative scale. Development of more specialized techniques such as isoelectric focusing, preparative electrophoresis, and affinity chromatography in the last 20 years has proven valuable for separation of many enzymes. These do not, however, always allow purification on a preparative scale with retention of biological activity and are not always applicable to a given protein. The introduction of suitable supports for high-performance liquid chromatography of proteins either by size exclusion, ion exchange, affinity, or reverse phase separations is allowing an advance in enzyme purification similar to the initial introduction of column supports in the 1950s. This chapter will focus on the application of high-performance ion exchange chromatography (HPIEC) for preparative scale purification of enzymes. The types of supports available will be discussed and examples from the literature given. Preparative separations of a number of proteins studied in this laboratory will be discussed and suggestions for general application of the technique detailed.

HPIEC SUPPORTS

A number of criteria have been suggested for development or choice of HPIEC support materials (Regnier, 1982; Hearn *et al.*, 1982). These include: (1) mechanical stability to a high flow rate; (2) availability in 5–10 μm particle size; (3) spherical shape; (4) high degree of hydrophilicity in order to minimize hydrophobic interactions between solutes and support; (5) high ion exchange capacity; (6) chemical stability particularly to mobile phases; (7) pore volume between 0.5 and 1.0 ml/g; (8) pore diameter from 300 to 1000 Å or higher; (9) ability to selectively bind proteins and allow high recoveries. In addition, the material should be readily available, easily packed and inexpensive. A number of support materials are currently available from numerous manufacturers. These materials meet many of the criteria listed above; however, none meet all of them. A relatively current list of commercially available ion exchange supports is provided in a recent review (Hearn *et al.*, 1982).

HPIEC support materials are of three types: (a) organic resins, (b) composite organic—inorganic supports, and (c) surface modified inorganic materials.

The first rigid organic supports for HPLC of proteins (Spheron, marketed by Lachema, Czechoslovakia) were developed by Mikes *et al.* (1976, 1978, 1981). They prepared polymethacrylate-based resins derivatized with either a weak anion exchange group, diethylaminoethyl (DEAE), or a carboxymethyl (CM) cation exchange group. Since derivatized Spheron is not available in a particle diameter less than 20 μm, its loading capacity is considerably lower than that of microparticulate resins. Pharmacia Fine Chemicals produces two organic-based HPIEC supports: Mono Q is a strong anion exchanger with a quaternary amine bonded phase and Mono S is a sulfonic acid derivatized strong cation exchanger. Both are available in 10 μm size. Similar materials (DEAE-5PW and SP-5PW) are also available from Toyo Soda Manufacturing (Tokyo, Japan).

Organic-based supports are chemically stable over a wider pH range (pH 3—10) than silica-based materials, but the slightly hydrophobic properties of organic matrices can result in non-specific adsorption of biomolecules. Organic-based supports, such as Trisacryls, Bio-beads, Sephacryl, and Toyo Pearl, exhibit poor mechanical stability with upper pressure limits in the range of 100—300 psi; however, packing in wide, short columns minimizes this problem. Mono-beads from Pharmacia (Piscataway, N.J.) and the PW ion exchangers from Toyo Soda will withstand pressures up to 3000 psi.

Composite organic—inorganic packings were recently introduced by Vanacek and Regnier (1982). They prepared a weak anion exchange material by crosslinking polyethyleneimine (PEI) to silica with pore diameters of 1000—4000 Å. Polymer layers of 60—80 Å thickness were formed, creating a large surface area available for ion exchange. Large pore diameter supports with thick polymer layers have the permeability properties of macroporous supports and the ion exchange capacities approaching those of small pore diameter, high surface area supports. Derivatized silica-based supports can withstand pressures of several thousand psi.

A third, and currently the most widely used, type of HPIEC support is a surface modified material. Regnier and co-workers (for references see Regnier, 1982) have prepared a number of ion exchange resins by crosslinking a thin layer of polymer on the surface of controlled porosity inorganic supports. Several examples are given below.

Chang *et al.* (1976) prepared a DEAE ether derivatized, glycerylpropylsilyl-coated controlled porosity glass support (DEAE-Glycophase G). In 1979, Alpert and Regnier (1979) reported the preparation of a weak anion exchanger by crosslinking a thin layer of PEI on the surface of a controlled porosity inorganic support.

A thin layer of amine was adsorbed onto the inorganic surface. The layer was then crosslinked under conditions that retained the amine on the surface. Such self-assembling coatings are highly reproducible. Columns were operated in the pH range 2–9.2 with no change in efficiency throughout the lifetime of the column. This material is produced by SynChrom, Inc. (Linden, Ind.) as SynChropak AX and is available in a number of pore sizes. Pharmacia also provides the material as Polyanion SI.

Kato *et al.* (1983) have recently prepared a weak anion exchange support of DEAE derivatized hydroxylated polyether with pore diameter of 1000 Å (marketed as TSK-GEL IEX-645 DEAE by Toyo Soda). Protein separations using this new support show significantly higher resolution than separations on IEX-545-DEAE Sil which is a silica-based weak anion exchanger with pore diameter of approximately 250 Å.

Preparation of high-performance cation exchange chromatography supports has lagged behind the development of the anion exchange materials. Recently, Regnier and co-workers (Gupta *et al.*, 1983) reported the preparation of a carboxymethyl polyamide weak cation exchanger. Silylated silica of 300 Å pore diameter was treated with PEI to form a polymeric mesh on the silica surface. The residual amines were then derivatized with diglycolic anhydride to generate a stable covalently bonded polymeric cation exchange support. A carboxymethyl support of this kind, SynChropak CM 300, is available from SynChrom and other sources. Alpert (1983) has described a weak cation exchange support with a self-assembling coating of poly(aspartic acid) on aminopropyl silylated silica with a pore diameter of 330 Å. Hemoglobin ion exchange capacity for this support was several times higher than that of the carboxymethyl polyamide material described above. Modification of the procedure used by Alpert could lead to the incorporation of sulfonate or phosphonate residues into the resin coating producing strong cation exchange materials. The support described by Alpert is commercially available from Custom LC, Inc. (Houston, Tex.) under the name PolyCAT A and from Brownlee Labs (Santa Clara, Calif.) under the name Aquapore CX-300.

FACTORS INVOLVED IN SEPARATIONS

As indicated above, many parameters will affect separations in ion-exchange chromatography. Factors involving choice of supports and chromatography conditions will be discussed.

Particle size
Separations require the diffusion of solute into and out of pore matrices to allow interaction with the charge groups on the support. Use of microparticulate column materials, 5–10 μm diameter, results in reduction of diffusion distances. Since large biological molecules diffuse at a much slower rate than small molecules, decreasing diffusion distances will reduce separation times and enhance resolution.

Pore diameter
As indicated by its name, the process of ion-exchange chromatography involves interaction of the solute with the surface of the support. Since greater than 95%

of the surface area of a porous support is within the pore network, the ability of molecules to penetrate the pore matrix will affect mass transfer and loading capacity.

The available surface area (A_a) for ion exchange partitioning is expressed by the equation (Regnier, 1982)

$$A_a = A_o + K_a A_i$$

where A_o is the outer surface area of particles in a bed, A_i is the internal surface area of the pores and K_a is the fraction, ranging from 0 to 1, of internal surface available for ion exchange absorption. As molecular dimensions approach pore size, K_a approaches zero and severe limitations to mass transfer result with concomitant loss of resolution. However, as A_i is decreased, as with larger pore size, the loading capacity decreases. An effective ion exchange support will be a compromise between resolution and loading capacity. Ion exchange packings with pore diameters of 250–300 Å have been used to purify proteins with molecular weights as high as 100 000. Gooding and Schmuck (1983a, b) have reported purification of trypsin and chymotrypsin on a 300 Å pore diameter weak cation exchanger. Using a 250 × 4.1 mm analytical column, 10 mg of protein was loaded with no loss of resolution. Vanacek and Regnier (1980) showed a decrease in resolution of AMP–ADP and ADP–ATP pairs with increasing pore diameter from 100 to 500 Å on 5 cm SynChropak anion exchange columns. Using the same support, ovalbumin (mol. wt 45 000) showed highest resolution with a 300 Å pore. Proteins with molecular weights greater than 100 000 exhibit highest ion exchange capacities on packings with 1000–5000 Å pore diameter.

Column length
Column length has relatively little effect upon resolution of proteins when gradients are used. Vanacek and Regnier (1980) studied the resolving power of 4.1 mm diameter SynChropak AX 300 columns varying in length from 50 to 200 mm. They found that for each additional 50 mm in column length, resolution of commerical ovalbumin was increased by approximately 11%. However, simply extending the gradient time while holding other variables constant led to a 15% increase in resolution on a 50 mm column. Kato *et al.* (1982) report almost identical peak intervals and peak widths of ovalbumin on 6 mm diameter TSK-GEL IEX-545 DEAE SIL columns of 75 mm and 150 mm length.

For analytical work there are several advantages to using small columns. These include the ability to elute proteins in a small volume, lower operating pressures, longer column life, and less expensive columns. The primary disadvantage to the use of small columns is their lower capacity; the loading capacity of a 50 mm column is at least ten times less than that of a 250 mm column. Vanacek and Regnier (1980) reported overloading problems when 1 mg protein samples were applied to a 50 mm SynChropak AX 300 column. Use of small columns results in very small elution volumes relative to the dead volume of the pumping and detector systems. If the dead volume of the instrument is 3–5 ml and the total elution volume is 5 ml, the ability of the instrument to provide the desired gradient within a stipulated time is reduced. The development of microbore technology is leading to improvements in instrument design that should help to minimize dead volume and improve performance of short columns. Use of isocratic elutions avoids this difficulty.

Mobile phase velocity

In isocratic elution, as the mobile phase velocity increases, the resolution of proteins decreases (Kato *et al.*, 1982). Resolution becomes more sensitive to flow rate as molecule size increases. These effects are due to a decrease in time available for mass transfer into the pores of the support. Kato *et al.* (1982) studied the effect of flow rate on salt gradient elutions of ovalbumin. They found that in the case of a constant gradient volume, resolution decreased with increasing flow rate. Vanacek and Regnier (1980) reported a 155% increase in resolution of ovalbumin on a SynChropak AX 300 column when the flow rate was decreased from 2 to 0.5 ml/min while gradient volume was kept constant. In the case of constant gradient time, the resolution increased with increasing mobile phase velocities in the range 0.25–0.50 ml/min.

Mobile phases

Since many commercial column supports are silica-based, mobile phases used with such supports have an upper limit of pH 8.0 in order to avoid destruction of the matrix. Most column supports can tolerate a wide range of aqueous and organic mobile phases. However, higher concentrations of organic solvents are not well suited to protein purification where recovery of biological activity is desired.

Gradient elution based on pH and/or ionic strength changes is most often used in HPIEC. The composition of the mobile phase, the type of gradient employed and other operational parameters are determined experimentally. Knowledge of the behavior of a given protein on conventional ion exchange column chromatographic supports is helpful in the choice of mobile phase in HPIEC. Though differences in ionic strength required for elution, resolution and separation times will occur, elution patterns on both conventional and high-performance ion-exchange columns are usually similar.

Kopaciewicz *et al.* (1983) studied the influence of various mobile phase ions on the retention of ovalbumin and soybean trypsin inhibitor on a strong anion exchange column and cytochrome c and lysozyme on a strong cation exchange column. For the anion exchange column, both retention and column selectivity (expressed as the ratio of retention times of soybean trypsin inhibitor/ovalbumin) changed when the cation (Na^+) was held constant and the anion varied. When the anion (Cl^-) was held constant while varying the cation, only the relative retention times changed. The use of different cations on the cation column did not significantly change the lysozyme/cytochrome retention ratio but selectivity was different with lithium as compared to other cations. Varying the anions altered both retention times and retention ratio; however, the changes were significantly less than those observed with the anion column. Gooding and Schmuck (1983a, b) have found that retention times for ribonuclease on a CM-type cation exchanger are affected very little by any change in either eluting anions or cations. Conversely, lysozyme has significant changes in retention time when either anion or cation composition is changed. Alterations of ion exchange support selectivity by changing mobile phase anion or cation components could aid in the separations of some complex mixtures.

Recently two groups have reported purifications using HPIEC of membrane proteins solubilized by non-ionic detergents. Welling *et al.* (1983) have reported the purification of the fusion protein of Sendai virus on a 50 X 5 mm Mono Q anion exchange column eluted with 0.1% Triton X-100 in phosphate buffered saline. Lundahl *et al.* (1983) fractionated human red cell membrane proteins on a Mono Q column eluted with a salt gradient containing *n*-octyl-beta-D-glucopyranoside. Though considerable overlap of protein peaks was seen, separation of glucose transportase from most of the accompanying phospholipids was achieved.

Temperature effects

Most protein purifications involving conventional chromatographic methods are performed at subambient temperatures in order to enhance protein stability. The reduction in elution time afforded by HPLC has, in most reported protein purifications, alleviated the need to refrigerate columns. Vanacek and Regnier (1980) reported a decrease in resolution of ovalbumin on a SynChropak AX 300 column when the temperature was lowered from 25 to 4°C. They also noted a two-fold increase in column head pressure. The use of subambient temperatures should only be used when purifying particularly temperature-sensitive proteins.

Frolik *et al.* (1982) reported reduction in both resolution and retention time of a mixture of insulin, β-lactoglobin, and carbonic anhydrase on a Partisil-10-SCX column when the column temperature was raised from 25 to 55°C. Protein recovery was significantly increased at the elevated temperature. The effect of the higher temperature on recovery of biological activity was not discussed. Peak width was decreased at 55°C as was elution time. The use of elevated temperatures may be suited to analytical preparations where recovery of biological activity is not a requirement or to the purification of temperature-stable proteins.

In general, it may be stated that HPIEC protein separations at room temperature (*ca.* 25°C) are satisfactory. Use of either higher or lower column temperatures would be indicated when protein lability is a consideration or enhanced recovery of total protein and/or biological activity could be achieved.

Effect of pH

Since proteins contain acidic and basic residues, their net charge is pH dependent. Under acidic conditions, basic residues will be ionized imparting a net positive charge to the protein molecule. Ionization of carboxyl groups under basic conditions results in negatively charged species. At some intermediate pH the net charge will be zero and the protein is said to be at its isoelectric point (pI).

Retention on ion exchange columns is dependent upon the charge of the molecules being separated. The 'net charge' concept (Kopaciewicz *et al.*, 1983), predicts that proteins will not be retained on ion exchange columns at their pI but will be retained on anion exchange supports above their pI and will be retained below their pI on cation exchange columns. Kopaciewicz *et al.* (1983) report that of 14 proteins studied, 11 deviated from the behavior predicted by the 'net charge' concept. For example, β-lactamase was retained on a Mono Q anion exchange column at a pH about one unit below its pI and on a Mono S cation exchanger approximately 0.5 pH units above its pI. Stanton *et al.* (1983) report that maximum resolution of a mixture of standard proteins on a Mono Q column was achieved when eluent pH was approximately 0.5 pH units below the pI of the most basic component. In our laboratory we have found that the basic isozyme of adenylosuccinate synthetase from rat will not bind to a cation support at neutral pH, either HPIEC or conventional, even though its pI is 8.9.

Deviations from the 'net charge' concept may be explained by the asymmetric distribution of charges within the protein molecule (Kopaciewicz *et al.*, 1983; Regnier, 1983). Even though the overall charge of a protein is zero at its pI, there may be areas of localized charge which will effect the interaction of the protein with the column support. These interactions can permit separation of proteins with similar isoelectric points.

APPLICATIONS

A number of studies have been done that indicate successful separations of enzymes can be achieved on HPIEC. Literature examples of such separations will be given in the first part of this section while a number of examples from our own laboratory will be discussed in the second part. The majority of the literature references involve more analytical uses and will not be emphasized here, but some analytical applications will be described to illustrate the utility of the technique.

A number of commercially available proteins have been chromatographed both individually and as mixtures on a variety of HPIEC supports. Lysozyme, cytochrome *c*, and hemoglobin have been the most used references for comparison of ion exchange capacity and resolution of column supports. Many other enzymes have been shown to be retained by HPIEC columns with subsequent elution with recovery of activity (for examples see Regnier, 1982).

Currently available HPIEC columns are replacing conventional ion exchange columns at all stages of purification schemes. This has resulted in a significant reduction in the time needed to elute the enzyme. A single DEAE column used in the purification of glucose oxidase required 36 h to run. This DEAE step was replaced with chromatography on a Mono Q column eluted with a 10 min gradient (Kirov *et al.*, 1983). Not only was run time drastically reduced but also the specific activity of the glucose oxidase preparation was 2.5-times higher than from the DEAE column.

Advantages other than the reduction of time required for the separations are being found. Rat brain hexokinase has been purified using a 250×4.1 mm SynChropak AX 300 column instead of a DEAE-cellulose step (Polakis and Wilson, 1982). Hexokinase is associated with the mitochondria and is easily solubilized with the addition of a low concentration of glucose-6-phosphate. DEAE-cellulose purified hexokinase loses the ability to rebind mitochondria, but hexokinase purified by HPIEC chromatography retains this characteristic. Also, the HPIEC column can separate the bindable form of the enzyme from the non-bindable form which results from DEAE-cellulose chromatography. The ability to purify bindable hexokinase will allow further studies on the reassociation of the enzyme to mitochondria.

One of the arguments against HPIEC for preparative work is that the column will not accept as large volumes or samples as conventional columns. The results of the study by Polakis and Wilson (1982) on the purification of hexokinase illustrates that volumes larger than the injection loop capacity can be loaded. Isolated mitochondria were treated with glucose-6-phosphate to solubilize hexokinase. The supernatant was concentrated to 100 ml, filtered, and pumped onto the SynChropak AX 300 column using a mini-pump followed by a wash with 100 ml of buffer. The column was then installed in an HPLC system and eluted with a 90 ml linear gradient. A single activity peak containing 6.3 mg of bindable hexokinase was eluted. Another example of preparative scale HPIEC purification is the study of beef pancreas asparagine synthetase (C. A. Luehr and S. M. Schuster, personal communication). The pooled protein from several affinity column runs was applied to a 250 × 10 mm SynChropak AX 300 column with a final yield of 6.5 mg of homogenous asparagine synthetase.

A commercial preparation of trypsin has been further purified with HPIEC (Gooding and Schmuck, 1983a, b). With a 250 × 10 mm SynChropak CM 300 column, 50 mg of trypsin could be bound and eluted with a sodium acetate gradient. The chymotrypsin contaminant normally found in trypsin preparations was separated. Since trypsin is prone to autolytic degradation during storage, the short time needed for HPIEC purification makes pure trypsin readily available.

The excellent resolution of HPIEC has been exploited to separate isozymes with a great deal of success. A major use has been in clinical applications, especially in serum analysis (Tomona et al., 1983; Kato et al., 1983; Vacik and Toren, 1982). Isozymes released upon tissue damage, such as release of lactate dehydrogenase isozymes in myocardial infarction, serve as diagnostic tools in medicine. The five different isozymes of lactate dehydrogenase can be separated from sera with a single column run requiring 20 min (Schlabach et al., 1980). The separation is so effective that multiplicity of the LD3 isozymes can be seen. The separation is achieved using a decreasing pH gradient starting at pH 7.8 on DEAE-glycophase packing material.

Another separation from plasma with clinical applications is the resolution of human hemoglobin types (Ou et al., 1983). Using the poly(aspartic acid)-silica column developed by Alpert (1983), separations of all the frequently encountered normal and variant hemoglobin types were accomplished. The procedure allows identification of twenty different hemoglobin types in less than 30 min. A Bis-Tris and KCN buffer (pH 6.5) is used with a salt and pH gradient to 0.2 M NaCl and pH 6.8. This separation supports the hypothesis that the separation by ion exchange columns is highly influenced by the protein configuration (Kopaciewicz et al., 1983; Regnier, 1983). Some of the Hb variants differ by only one amino acid, which is only a 1–2% difference in primary structure. This difference is not enough for electrophoresis to effect the separations achieved, supporting the idea that secondary or tertiary protein structure has a significant influence on the binding of proteins to ion exchange columns.

Binding of substrates to proteins may also cause changes in the tertiary structure of proteins, which can alter the retention time of the protein. The binding of iron to the human plasma protein transferrin illustrates this point (Strahler *et al.*, 1983). Transferrin contains two iron binding sites, one near the amino terminus and the other at the carboxy terminus. By loading plasma onto a SynChropak AX 300 column and eluting with a multi-stage gradient, the four possible forms of transferrin are fully resolved. Rechromatographs of each form subsequently saturated with iron showed a single peak corresponding to the fully occupied transferrin.

To demonstrate the power of HPIEC technique as a preparative tool, a purification of glucose-6-phosphate dehydrogenase from a yeast enzyme concentrate has been studied (Lindblom, 1983). The entire purification scheme consists of two HPIEC columns. A large column (100 × 10 mm) packed with Polyanion SI-17 μm was used in the first step followed by elution from a 50 × 5 mm column of the same support but with smaller pore size. More than 100 mg of sample could be loaded on the larger column without broadening of the activity peak. The active fractions were pooled and diluted to achieve the same ionic strength as the starting buffer of the second column. A 10 ml injection loop was used to apply the sample onto the second column. A NaBr gradient was used to elute both columns. Polyacrylamide gel electrophoresis of the purified protein showed a single band. The recovery of activity was 85%. The combined time for both column gradients was only 95 min.

The short time required to elute these columns accentuates their usefulness. Short gradient times of 20–60 min allow considerable fine tuning of solvent composition, gradient shape and flow rate. The small sample size required allows optimization even when starting materials are in short supply. In development of the purification of glucose-6-phosphate dehydrogenase described above, 10 μl samples of the yeast concentrate were applied to the 50 × 5 mm column for optimization of the gradient conditions. From these studies a five-stage gradient was chosen. The glucose-6-phosphate dehydrogenase was eluted in a relatively steep portion of the gradient. A shallow gradient prior to elution and a further decrease in the slope of the gradient as the glucose-6-phosphate dehydrogenase eluted helped to increase the resolution.

Many gradient alterations are possible with the HPLC equipment available. A step gradient can be used when a linear gradient fails to resolve a group of peaks. When the step gradient was utilized for separation of crude human pituitary fractions (Stanton *et al.*, 1983) a group of unresolved peaks was separated into distinct fractions at each step in the gradient. This technique may be useful in preparative application for obtaining fractions of a crude sample which can be resolved in a subsequent purification step. The step gradient may also be used as a primary tool to develop conditions for eluting a new protein.

In our laboratory a number of enzymes have been purified preparatively using HPIEC. In one of our initial studies we were able to purify the acidic isozyme of adenylosuccinate synthetase from rat liver (Rudolph and Clark, 1982). In these studies the enzyme was initially purified by conventional techniques including salt fractionation, DEAE-cellulose chromatography, and an affinity column. At this stage of the purification the enzyme was extremely labile and any conventional separations led to complete loss of activity. The enzyme was chromatographed on a SynChropak AX 300 column (250 × 4.6 mm) at room temperature with phosphate buffer (pH 7.5) containing dithiothreitol and EDTA. Two consecutive runs yielded an essentially homogeneous preparation of the enzyme with a specific activity of 3000 U/mg of protein (Rudolph and Clark, 1982) with nearly quantitative recovery of activity. In more recent studies in our laboratory, we have been able to obtain specific activities of over 5000 U/mg which represents a pure preparation of the protein. Using the 250 × 4.6 mm column we have applied over 60 mg of an impure preparation with retention of most of the protein. Characteristics of the initially eluted bands indicated that this protein load exceeded the capacity of the column. In subsequent studies we have used a 250 × 10 mm AX 300 column with a proportional increase in capacity without loss of resolution. This particular isozyme of adenylosuccinate synthetase had not been purified to homogeneity previously due to its lability when nearly pure. The fact that good separation could be achieved in 30 min or less combined with the ability to make whatever additions were necessary to the buffer to prevent inactivation was the key to this purification.

In a collaborative study with Diane Ingolia and Dr Rodney Kellems of the Baylor College of Medicine, we have developed a purification scheme for adenosine deaminase from a gene amplified cell line (Yueng et al., 1983). With the use of growth conditions which included specific inhibitors, cells were obtained in which adenosine deaminase represented about 50% of the soluble protein. We have been able to purify the enzyme to essential homogeneity in a single step by applying the soluble cell extract to a SynChropak AX 300 column as shown in Fig. 1. The activity peak coincides with the protein peak at 35–38 min. The protein across this peak exhibits only one band on SDS gels suggesting that the shoulder on the peak may represent a slightly modified enzyme. We have been able to load up to 15 mg of extract on a 250 × 46 mm column without loss of resolution and have used up to 50 mg on a 250 × 10 mm column. The shoulder observed in Fig. 1 is consistent regardless of the amount of enzyme applied to the column. Since adenosine deaminase represents about 50% of the protein present, this indicates

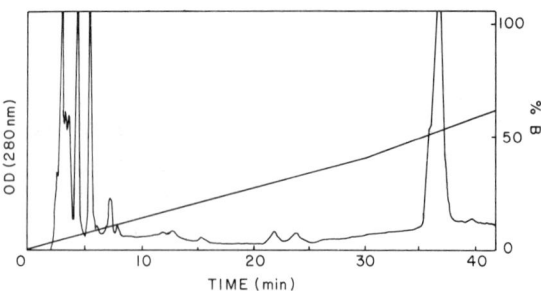

Figure 1. Purification of adenosine deaminase from BO-1/50 cells on a 250 × 4.1 mm SynChropak AX 300 column. One hundred μl of a centrifuged cell extract was applied. The flow rate was 1 ml/min with detection at 280 nm. The initial buffer (A) was 5 mM potassium phosphate, pH 7.5 with 1 mM dithiothreitol. Buffer B was 100 mM potassium phosphate, pH 7.5 with 1 mM dithiothreitol. The gradient (%B) is indicated by the series of straight lines.

a capacity of about 7 mg of adenosine deaminase on the smaller column. The capacity of the bigger column should be about 35 mg of the deaminase. This capacity is certainly sufficient to allow large scale purification of the enzyme in a short time. A short SynChropak ASC guard column was used along with a SynChrosorb solvent conditioning column. Even with direct application of the soluble cell extract we have found column life to be quite long, often using the same column for over one year.

Another application of preparative HPIEC that we have developed has been for final purification of proteins prior to crystallization. In collaboration with Dr Florante Quiocho and his co-workers at Rice University we have taken apparently homogenous preparations of a number of proteins being studied in his laboratory and chromatographed them on SynChropak AX columns. In all cases the proteins were found to have contaminants that were removed by the chromatographic precedure. It appears that this final purification will allow growth of larger crystals of the proteins studied. Two examples of such separations will be given. The first involves sulfate-binding protein from *Salmonella typhimurium* that was purified as previously described (Pardee, 1966). The previous purification scheme as modified in Dr Quiocho's laboratory used DE-53 chromatography as a final step, giving apparently homogenous protein as evidenced by gel electrophoresis. The result of chromatographing sulfate-binding protein which has sulfate bound to the protein is shown in Fig. 2. Two to three contaminants were removed by this procedure. Similar separations have been done with arabinose, maltose and leucine-isoleucine-valine binding proteins.

In another study with Dr Quiocho, we have chromatographed the phage P22 tail protein (Sauer *et al.*, 1982). The purified protein was supplied by Dr Peter Berget from the University of Texas Medical School at Houston and then purified further in Dr Quiocho's laboratory by two chromatography steps on DE-53 cellulose. The protein eluted from the second DE-53 column as one symmetrical peak and only one protein band was observed on electrophoresis gels. The results of chromatography of 9.2 mg of the P22 tail protein on a 250 × 10 mm SynChropak AX 1000 column are shown in Fig. 3. When a very shallow step gradient was utilized, the resultant protein peaks were not symmetrical, indicating that the applied protein was not homogeneous. The protein has a molecular weight of about 210 000

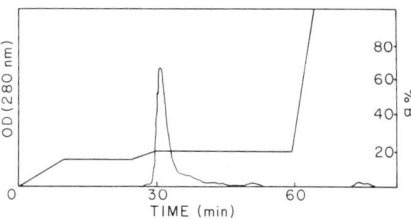

Figure 2. Purification of sulfate binding protein from *Salmonella typhimurium* on a 250 × 10 mm SynChropak AX 300 column. A 6 mg sample was applied. The flow rate was 2 ml/min with detection at 280 nm. The initial buffer (A) was 1 mM potassium phosphate, pH 7.6 with 0.02% azide. Buffer B contained 200 mM sodium phosphate in buffer A. The gradient (%B) is indicated by the series of straight lines.

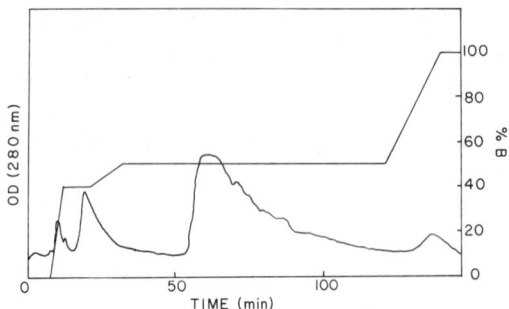

Figure 3. Purification of *Salmonella* phage P22 tail protein on a 250 × 4.6 mm SynChropak AX 1000 column. A 9.2 mg sample was applied. The flow rate was 1 ml/min with detection at 280 nm. The initial buffer (A) was 1 mM potassium phosphate, pH 7.6 with 0.02% azide. Buffer B contained 500 mM potassium chloride in buffer A. The gradient (%B) is indicated by the series of straight lines.

and chromatographed well on the 1000 Å support but could not be recovered from the 300 Å support. We have found similar results for aspartase which has a similar molecular weight.

Other enzymes that have been purified on HPIEC in our laboratory include yeast hexokinase, chick muscle adenylosuccinate lyase, and aspartase from *Escherichia coli*.

DEVELOPMENT OF PURIFICATION PROCEDURES USING HPIEC

The use of HPIEC in purification procedures requires no more effort than conventional chromatographic procedures. The equipment investment can be minimal and detection procedures are nearly the same as for conventional chromatography. For larger bore columns (10 mm or wider) the backpressure is usually quite low so that the column could be run with a peristaltic pump. In our laboratory we use systems ranging from a simple single piston pump with bottle gradients to a microprocessor-controlled dual pump system. The application of the sample onto the column can be a problem. Since sample size is not a consideration if the desired protein binds to the column, rather large volumes will be commonly applied to the column. This can be accomplished with a peristaltic pump if the backpressure is low. Proteins can be monitored by simple flow through optical detectors or the absorbance of fractions can be determined in a spectrophotometer and the activity of an enzyme assayed. Post-column reactors have been developed for various enzymes. These are certainly convenient but are more applicable for routine analytical work than for purification purposes.

An example of full utilization of available technology is illustrated by a recent study (Lindblom and Fagerstam, 1983). A closed HPIEC system was used to purify seven enzymes from chicken breast muscle. The entire separation was controlled by a microprocessor and consisted of multiple desalting, anion exchange and cation exchange columns. Motor driven multiport valves were used to direct the solute and solvent flow. This type of system could be readily scaled up to allow unattended purification of proteins from large-scale sources such as continuous fermentations.

In designing purification the choice of column will depend on the separation desired. We have used the SynChropak AX and CM columns with great success but the supports available from other sources should be equally useful in this type of work. The physical characteristics of the supports need to be considered when choosing a particular column.

A guard column with support material similar to the column should be used to prevent irreversible binding and clogging of the column. In addition, a silica precolumn should be included prior to where the sample is injected when a silica-based support is used with aqueous solvents. The enzyme solutions should be passed through a membrane-type filter or centrifuged prior to injection onto the column. These precautions and proper storage practices should allow the column to be used for many separations making the procedure very cost effective. The use of metal columns and tubing has been criticized but we have not experienced difficulty in this regard. Inclusion of chelators and other factors such as sulfhydryl compounds should usually be sufficient to preserve enzyme activity.

In addition, the point in the purification scheme at which HPIEC is employed will also affect column life. It would be preferable if initial conventional separation steps were done prior to HPIEC. A suitable design for developing a new purification scheme would involve an initial concentration step such as ammonium sulfate fractionation followed by a molecular sieve column and then a conventional ion exchange cellulose or other matrix column. These steps would remove material that would bind very tightly to the HPIEC support and provide initial purification so that more efficient separations could be achieved. Another option would involve an initial affinity column followed by HPIEC. The examples described above, both from our laboratory and others, indicate that many combinations are possible. The speed and resolving power of this technique make optimization of column purification procedures possible. One important point to note when developing these procedures is that due to its high ion exchange capacity a column can take considerable time to equilibrate when using a dilute buffer. The pH and conductivity of the effluent should be monitored. A procedure that we have been using recently involves elution with a very shallow gradient or isocratic conditions. With conventional columns such a procedure would hopelessly dilute the protein but we have effected some very specific separations with little dilution using the HPIEC supports.

FUTURE APPLICATIONS AND POTENTIAL

High-performance ion-exchange chromatography has not been utilized as frequently for protein purification as molecular sieve and reverse phase methods but the recent results from our laboratory and others suggest that for preparative separation of biologically active molecules it may be the method of choice. Previous experience with ion exchange on gel or cellulose supports will generally form a basis for design of HPLC separations. This allows development of a separation with minimal effort.

In addition, the technique can easily be scaled up for industrial applications with the only limitation being the cost of the support. For many purposes a 5 μm support is not required so that the cost factor can be reduced with use of larger supports. In addition, the functional capacity of the supports is sufficiently large that less material would be needed than with conventional chromatography supports. With proper use the lifetime of the columns is also quite long and the ability to recycle *in situ* reduces operational downtime and makes the separations more efficient in terms of time and labor.

Acknowledgements
We wish to thank Brian Kubena and Diane Ingolia for providing some of the data described in this review and Dr Andrew J. Alpert for a critical review of the manuscript. This research was supported by grant 14030 from the National Cancer Institute and grant C-582 from the Robert A. Welch Foundation. BFC and JG have been recipients of predoctoral fellowships from the Robert A. Welch Foundation.

REFERENCES

Alpert, A. J. (1983). Cation-exchange high-performance liquid chromatography of proteins on poly(aspartic acid)-silica. *J. Chromatogr. 266*, 23–37.

Alpert, A. J. and Regnier, F. E. (1979). Preparation of a porous microparticulate anion-exchange chromatography support for proteins. *J. Chromatogr. 185*, 375–392.

Chang, S., Gooding, K. M., and Regnier, F. E. (1976). High-performance liquid chromatography of proteins. *J. Chromatogr. 125*, 103–114.

Frolik, C. A., Dart, L. L., and Sporn, M. B. (1982). Variables in the high-pressure cation-exchange chromatography of proteins. *Anal. Biochem. 125*, 203–209.

Gooding, K. M. and Schmuck, M. N. (1983a). Ion selectivity in the high performance cation-exchange chromatography of proteins. *3rd International Symposium on HPLC of Proteins, Peptides and Polynucleotides*. Abstract 707.

Gooding, K. M. and Schmuck, M. N. (1983b). Purification of trypsin and other basic proteins by high-performance cation-exchange chromatography. *J. Chromatogr. 266*, 633–642.

Gupta, S., Pfannkoch, E., and Regnier, F. E. (1983). High-performance cation-exchange chromatography of proteins. *Anal. Biochem. 128*, 196–201.

Hearn, T. W., Regnier, F. E., and Wehr, C. T. (1982). HPLC of peptides and proteins. *American Laboratory*, October, 18–39.

Kato, Y., Komiya, K., and Hishimoto, T. (1982). Study of experimental conditions in high-performance ion-exchange chromatography of proteins. *J. Chromatogr. 246*, 13–22.

Kato, Y., Nakanmura, K., and Hashimoto, T. (1983). New ion exchanger for the separation of proteins and nucleic acids. *J. Chromatogr. 266*, 385–394.

Kirov, M. S., Taseva, D. K., and Ivanova, I. V. (1983). Purification of glucose oxidase from *Pen. chrisogenum*. *3rd International Symposium on HPLC of Proteins, Peptides and Polynucleotides*. Abstract 217.

Kopaciewicz, W., Rounds, M. A., Fausnaugh, J., and Regnier, F. E. (1983). Retention model for high-performance ion-exchange chromatography. *J. Chromatogr. 266*, 3–21.

Lindblom, H. (1983). Rapid chromatographic method for the isolation of glucose-6-phosphate dehydrogenase from yeast enzyme concentrate. *J. Chromatogr. 266*, 265–271.

Lindblom, H. and Fagerstam, L. G. (1983). An automatic multidimensional chromatography system for the separation of proteins. *3rd International Symposium on HPLC of Proteins, Peptides and Polynucleotides*. Abstract 810.

Lundahl, P., Lindblom, H., and Fagerstam, L. G. (1983). Fractionation of human red cell membrane proteins by ion exchange chromatography in detergent on Mono-Q. *3rd International Symposium on HPLC of Proteins, Peptides and Polynucleotides.* Abstract 707.

Mikes, O., Strop, P., Zbrozek, J., and Coupek, J. (1976). Chromatography of biopolymers and their fragments on ion exchange derivatives of the hydrophilic macroporous synthetic gel Spheron. *J. Chromatogr. 119*, 339–354.

Mikes, O., Strop, P., and Sedlackova, J. (1978). Rapid chromatographic separation of technical enzymes on Spheron ion exchangers. *J. Chromatogr. 148*, 237–245.

Mikes, O., Sedlackova, J., Rexova-Benkova, L., and Omelkova, J. (1981). High-performance liquid chromatography of pectic enzymes. *J. Chromatogr. 207*, 99–114.

Ou, C.-N., Buffone, G. J., Reimer, G. L., and Alpert, A. J. (1983). High-performance liquid chromatography of human hemoglobins on a new cation exchanger. *J. Chromatogr. 266*, 197–205.

Pardee, A. B. (1966). Purified sulfate binding protein. *J. Biol. Chem. 241*, 5886–5892.

Polakis, P. G. and Wilson, J. E. (1982). Purification of highly bindable rat brain hexokinase by high performance liquid chromatography. *Biochem. Biophys. Res. Comm. 107*, 937–943.

Regnier, F. E. (1982). High-performance ion-exchange chromatography of proteins: the current status. *Anal. Biochem. 126*, 1–7.

Regnier, F. E. (1983). Surface mediated separations of proteins. The importance of three-dimensional structure. *Liquid Chromatogr. 1*, 350–352.

Rudolph, F. B. and Clark, S. W. (1982). High performance liquid chromatography of proteins: purification of the acidic isozyme of adenylosuccinate synthetase from rat liver. *Anal. Biochem. 127*, 193–197.

Sauer, R. T., Krovatin, W., Poteete, A. R., and Berget, P. B. (1982). Phage p22 tail protein: gene and amino acid sequence. *Biochemistry 21*, 5811–5815.

Schlabach, T. O., Fulton, J. A., Mockridge, P. B., and Toren, E. C. (1980). Determination of serum isozyme activity profiles by HPLC. *Anal. Chem. 52*, 729–733.

Stanton, P. G., Simpson, R. J., Lambrou, F., and Hearn, M. T. W. (1983). High performance liquid chromatography of amino acids, peptides and proteins. *J. Chromatogr. 266*, 273–279.

Strahler, J. R., Rosenblum, B. B., Hanash, S., and Butkunas, R. (1983). Separation of transferrin types in human plasma by anion-exchange high-performance liquid chromatography. *J. Chromatogr. 266*, 281–291.

Tomono, T., Ikeda, H., and Tokunaga, E. (1983). High-performance ion-exchange chromatography of plasma proteins. *J. Chromatogr. 266*, 39–47.

Vacik, D. N. and Toren, E. C. (1982). Separation and measurement of isozymes and other proteins by high performance liquid chromatography. *J. Chromatogr. 228*, 1–31.

Vanacek, G. and Regnier, F. E. (1980). Variables in the high-performance anion-exchange chromatography of proteins. *Anal. Biochem. 109*, 345–353.

Vanacek, G. and Regnier, F. E. (1982). Macroporous high-performance anion-exchange supports for proteins. *Anal. Biochem. 121*, 156–169.

Welling, G. W., Groen, G., and Welling-Webster, S. (1983). Isolation of Sendai virus f protein by anion-exchange high-performance liquid chromatography in the presence of Triton X-100. *J. Chromatogr. 266*, 629–632.

Yueng, C.-Y., Ingolia, D. E., Bobonis, C., Dunbar, B. S., Riser, M. E., Siciliano, M. J., and Kellems, R. E. (1983). Selective overproduction of adenosine deaminase. *J. Biol. Chem. 258*, 8338–8345.

Progress in HPLC, Vol. 1, pp. 149–155
Parvez *et al.* (Eds)
© 1985 VNU Science Press

Application of HPLC for analysis and purification of catecholamine-synthesizing enzymes

K. KOJIMA,[1] S. PARVEZ,[2] H. PARVEZ,[2] Y. KATO,[3] and T. NAGATSU[1]

[1] Laboratory of Cell Physiology, Department of Life Chemistry, Graduate School at Nagatsuta, Tokyo Institute of Technology, Yokohama, Japan and
[2] Neuropharmacology Unit, University of Paris XI, 91405 Orsay, France and
[3] Central Research Laboratory, Toyo Soda Mfg Co. Ltd, Yamaguchi, Japan

INTRODUCTION

Enzymes of catecholamine biosynthesis, tyrosine hydroxylase (TH), aromatic L-amino acid decarboxylase (AADC), dopamine β-hydroxylase (DBH), and phenylethanolamine N-methyltransferase (PNMT), have been extensively studied in biochemistry, pharmacology, physiology, and neurobiology. These enzymes have been purified to homogeneity using conventional column chromatographies. However, these enzymes are difficult to purify because of their instability, low tissue concentration, or multiplicity.

High-performance liquid chromatography (HPLC) has been extensively developed in the last ten years, and great advances have also been made in the study of enzymes and proteins by HPLC. One application of HPLC for enzymes is the assay of enzyme activity, and the other is the purification of enzyme proteins. The application of HPLC to the assay of enzyme activity has become one of the most important and valuable techniques in enzymology. The assay for catecholamine-metabolizing enzymes has been recently summarized (Nagatsu and Kojima, 1984).

Purification of enzyme proteins by HPLC has been recently introduced. Several techniques have been developed for separation and isolation of proteins by HPLC in the size exclusion or gel filtration (GF-) (Furuta *et al.* 1981; Nelson and Traub, 1982; Finlay *et al.*, 1982; Hurley and Stryer, 1982; Lloyd and Walega, 1981), affinity (AF-), ion exchange (IE-) (Gooding *et al.*, 1979; Rudolph and Clark, 1982; Ansari *et al.*, 1983) and reverse-phase (RP-) modes (Congote *et al.*, 1979; Lewis *et al.*, 1980; Kilpatrick *et al.*, 1982; Parman and Rideout, 1983). However, applications of HPLC to larger proteins such as enzymes are still rather scarce. The application of HPLC to larger proteins is most developed in the field of GF-chromatography (or gel-permeation chromatography, GPC). Proper combination of chromatographic conditions, such as the support, eluting solvents, and flow rate must be chosen for the best utilization of GPC to protein purification. For enzyme purification, more careful treatments in HPLC are needed as compared to non-enzymatic proteins. For example, temperature must be lower for enzyme proteins than for non-enzymatic proteins, and the mobile phase should not decrease the enzyme activity or denature the enzyme molecule. Among catecholamine-related enzymes, only TH and monoamine oxidase type B have been tried to purify using a HPLC step. TH of rabbit adrenals was subjected on GF-HPLC (Lloyd and Walega, 1981) and monoamine oxidase (type B) of human platelet on anion exchange HPLC (Ansari et al., 1983).

In this chapter, we describe the isolation and partial purification of catecholamine-synthesizing enzymes by HPLC.

METHODS

Assay of enzyme activity and enzyme protein

TH activity was assayed by the HPLC method of Oka et al. (1982). DBH activity was determined based on the HPLC method of Matsui et al. (1982). Other enzymes, AADC (Nagatsu et al., 1984) and PNMT (Trocewicz et al., 1982) were also assayed using the HPLC methods. Dihydropteridine reductase (DPR) activity was assayed spectrophotometrically as described by Togari et al. (1983).

TH and DBH proteins were measured by enzyme immunoassays (Mogi et al., 1984).

Methods in HPLC

HPLC were performed either by a Toyo Soda HLC-803D system with columns of TSK-Gel SW types or by a Gilson system 42 gradient HPLC with a model 111 detector and a SynChropak AX300 column. Few conventional column chromatographies were also used for separation.

RESULTS AND DISCUSSION

Separation of catecholamine-synthesizing enzymes by HPLC

First, gel filtration (GF-) [or gel permeation (GP-)] HPLC was tried to separate standard proteins (Table 1) with M_r from 669 000 to 12 400. Table 2 shows the M_r of catecholamine-synthesizing enzymes which we studied. Combination of one TSK G2000SW column and one TSK G3000SW column or two TSK G3000SW columns did not result in good separation of proteins with medium M_r. Combination of one TSK G2000SW column, one TSK G3000SW column and one TSK G4000SW column or combination of two TSK G3000SW columns and one TSK G4000SW column gave a good separation for our purpose.

Figure 1 shows the calibration curve of GF-HPLC (two TSK G3000SW columns + one TSK G4000SW column) for standard proteins.

In this HPLC system, four catecholamine-synthesizing enzymes (TH, AADC, DBH, and PNMT) were almost completely separated with apparent M_r which agree with the reported values. Figure 2 shows the separation pattern of catecholamine-synthesizing enzymes in 100 000g supernatant from chromaffin granules of bovine adrenal medulla. Addition of 0.2 M NaCl resulted in better separation of

Table 1. List of standard proteins for GF-HPLC (GPC)

Protein	Molecular weight	Supplier*
Blue dextran	2 000 000	P
Thyroglobulin (bovine thyroid)	669 000	P
Ferritin (horse spleen)	450 000	B
	440 000	P
Glutamate dehydrogenase (yeast)	290 000	O
Catalase (bovine liver)	240 000	B
	232 000	P
Aldolase (rabbit muscle)	158 000	B, P
Lactate dehydrogenase (porcine heart muscle)	142 000	O
Albumin (bovine serum)	68 000	B
	67 000	P
Enolase (yeast)	67 000	O
Ovalbumin (hen egg)	45 000	B
	43 000	P
Adenylate kinase (yeast)	32 000	O
Chymotrypsinogen A (bovine pancreas)	25 000	B, P
Ribonuclease-A (bovine pancreas)	13 700	P
Cytochrome c	12 500	B
(horse heart muscle)	12 400	P
	12 400	O
Tyrosine	181	S

*B Boehringer Mannheim GmbH, Mannheim, FRG; P Pharmacia Fine Chemicals AB, Uppsala, Sweden; O Oriental Yeast Co., Ltd, Tokyo, Japan; S Sigma Chemical Co., St Louis, Mo., USA.

Table 2. Molecular weight of catecholamine-synthesizing enzymes as measured by GF-HPLC (GPC)

Enzyme	Molecular weight	Reference
Dopamine-β-hydroxylase (human adrenal)	350 000	Kojima *et al.* (unpublished)
Tyrosine hydroxylase (human adrenal)	280 000	Kojima *et al.* (1984)
Aromatic L-amino acid decarboxylase (human lung)	100 000	Ichinose *et al.* (unpublished)
Dihydropteridine reductase (human liver)	47 500	Firgaira *et al.* (1981)
Phenylethanolamine N-methyl-transferase (bovine adrenal)	38 000	Connett and Kirshner (1970)

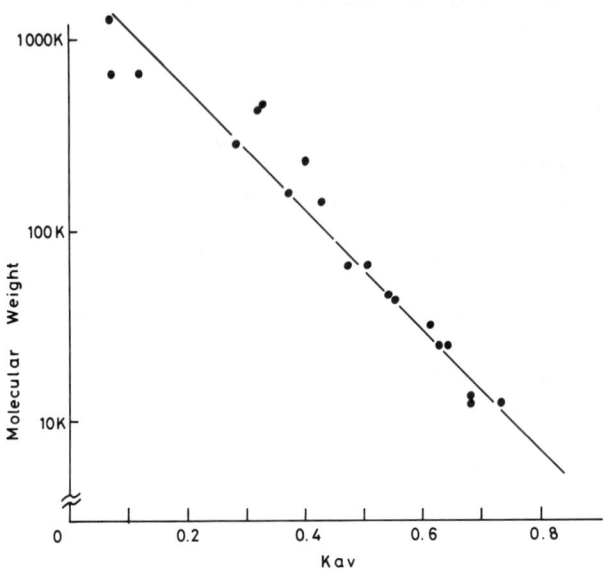

Figure 1. Calibration curve of a HPLC system for proteins with two TSK G3000SW and one TSK G4000SW columns. Mobile phase: 0.2 M KPB (pH 6.8) containing 0.2 M NaCl and 1 mM DTT. Temperature: 4°C. Flow rate: 0.5 ml/min.

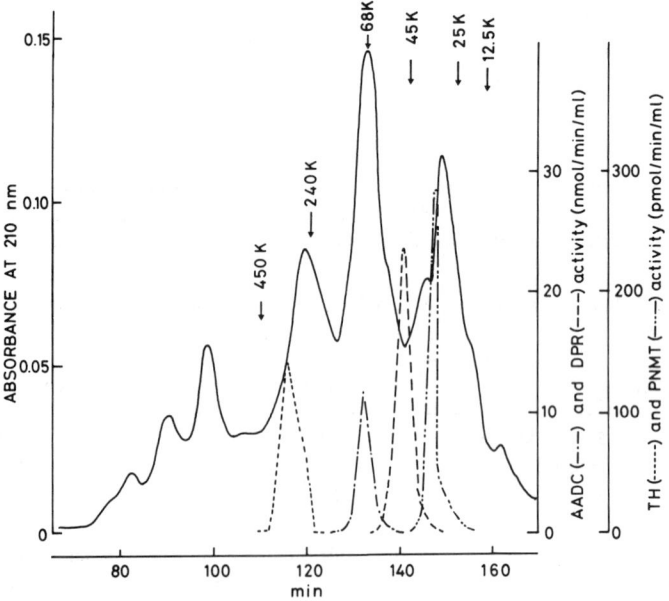

Figure 2. HPLC Chromatogram of the 100 000g supernatant of bovine adrenal medullary granules with four enzymes activities. Column: Two TSK G3000SW and one TSK G4000SW. Mobile phase: 0.2 M KPB (pH 6.8) containing 0.2 M NaCl and 1 mM DTT. Flow rate: 0.4 ml/min.

four catecholamine-synthesizing enzymes, although salt did not improve the separation of several standard proteins.

Figure 3 shows the separation of TH and DBH using the detection by sandwich EIA for the enzyme protein of TH and enzyme activity of DBH. Enzyme proteins were completely separated by this HPLC system, and the enzyme activities coincided with the enzyme proteins detected by EIA.

This GF-HPLC method takes 20–60 min (from void volume to salt volume) for one column and 60–150 min for three columns at 0.5 ml/min of flow rate. These results suggest that GF-HPLC gives sharp and rapid separation depending upon their molecular weights.

The combination of these GF columns is useful for both analytical and preparative systems.

IE-HPLC, anion exchange type in this case, is also useful for analytical purpose. This IE-HPLC could separate effectively active and less active form of TH (Mogi *et al.*, 1984).

Purification of aromatic L-amino acid decarboxylase (AADC) by HPLC

When highly sensitive assay for enzyme activity or protein is available, analytical columns for HPLC are enough for the micropurification.

Enzyme purification can be done by the combination of conventional chromatographies and analytical HPLC. AADC is subjected to this purification procedure (Kojima *et al.*, 1984). Table 3 shows the summary of purification of AADC from bovine adrenal medulla. The separation pattern by HPLC was better than that by conventional GF- and IE-columns.

One disadvantage is that the amount of sample subjected to a HPLC column is smaller than that of a conventional column. HPLC for enzyme purification could be used with an autosampler, a fraction collector, and a cooling system.

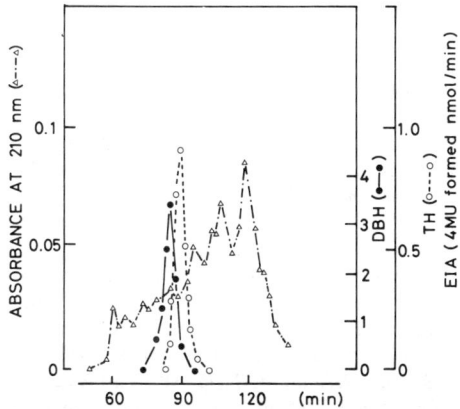

Figure 3. HPLC Chromatogram of the 100 000*g* supernatant of human adrenals with TH and DBH profiles. Columns: Two TSK G3000SW and a TSK G4000SW. Mobile phase: 0.2 M KPB (pH 6.8) containing 1 mM DTT. Temperature 4°C. Flow rate: 0.5 ml/min. 4MU: 4-methylumbelliferone.

Table 3. Purification of aromatic L-amino acid decarboxylase (AADC) of bovine adrenal medulla by HPLC

Step	Volume (ml)	Protein (mg)	Specific activity (nmol/min/mg protein)	Total activity (nmol/min)	Purification (*n*-fold)	Yield (%)
100 000*g* supernatant	6.0	38.6	32.1	1240	1	100
HPLC-DEAE	56.3	2.07	305	631	9.5	51
GPC	6.5	0.018	1800	32.4	56.1	2.6

However, ten- to hundred-fold higher amounts of an enzyme can be applied for a TSK preparative column of SWG types. Preparative type of IE-column is also available. Both steps of HPLC were quite effective for AADC purification.

CONCLUSIONS

The application of HPLC for enzyme purification is now developing rapidly. Although the experiments shown here are only a few examples, the results prove that HPLC is quite useful for analysis and purification of catecholamine-related enzymes.

Acknowledgements
The authors deeply thank Mr M. Mogi, Mr H. Inagaki and Mr H. Ichinose (Laboratory of Cell Physiology, Department of Life Chemistry, Graduate School at Nagatsuta, Tokyo Institute of Technology) for their valuable assistance in the experiments.

REFERENCES

Ansari, G. A. S., Patel, N. T., Fritz, R. R., and Abell, C. W. (1983). Purification of human platelet monoamine oxidase B by high performance liquid chromatography. *J. Liq. Chromatogr. 6*, 1407–1419.

Congote, L. F. Bennett, H. P. J., and Solomon, S. (1979). Rapid separation of the α, β, Gγ and Aγ human globin chains by reversed-phase high pressure liquid chromatography. *Biochem. Biophys. Res. Commun. 89*, 851–858.

Connett, R. J. and Kirshner, N. (1970). Purification and properties of bovine phenylethanolamine N-methyltransferase. *J. Biol. Chem. 245*, 329–334.

Finlay, T. H., Katz, J., and Levitz, M. (1982). Purification and properties of an estrogen-stimulated hydrolase from mouse uterus. *J. Biol. Chem. 257*, 10914–10919.

Firgaira, F. A., Cotton, R. G. H., and Danks, D. M. (1981). Isolation and characterization of dihydropteridine reductase from human liver. *Biochem. J. 197*, 31–43.

Furuta, S., Miyazawa, S., and Hashimoto, T. (1981). Purification and properties of rat liver Acyl-CoA dehydrogenases and electron transfer flavoprotein. *J. Biochem. 90*, 1739–1750.

Gooding, K. M., Lu, K.-C., and Regnier, F. E. (1979). High-performance liquid chromatography of hemoglobins. I. Determination of hemoglobin A_2. *J. Chromatogr. 164*, 506–509.

Hurley, J. B. and Stryer, L. (1982). Purification and characterization of the γ-regulatory subunit of the cyclic GMP phosphodiesterase from retinal rod outer segments. *J. Biol. Chem. 257*, 11094–11099.

Kilpatrick, D. L., Jones, B. N., Lewis, R. V., Stern, A. S., Kojima, K., Shively, J. E., and Udenfriend, S. (1982). An 18 200-dalton adrenal protein that contains four [Met]enkephalin sequences. *Proc. Natl. Acad. Sci. USA 79*, 3057–3061.

Kojima, K., Mogi, M., Oka, K., and Nagatsu, T. (1984). Purification and immunochemical characterization of human adrenal tyrosine hydroxylase. *Neurochem. Int. 6*, 475–480.

Lewis, R. V., Stern, A. S., Kimura, S., Stein, S., and Udenfriend, S. (1980). Enkephalin biosynthetic pathway: Proteins of 8000 and 14 000 daltons in bovine adrenal medulla. *Proc. Natl. Acad. Sci. USA 77*, 5018–5020.

Lloyd, T. and Walega, M. A. (1981). Purification of tyrosine hydroxylase by high-pressure liquid chromatography. *Anal. Biochem. 116*, 559–563.

Matsui, H. Yamamoto, C., and Nagatsu, T. (1982). Purification and properties of bovine brain dopamine β-hydroxylase. *J. Neurochem. 39*, 1066–1071.

Mogi, M., Kojima, K., and Nagatwu, T. (1984). Detection of inactive or less active forms of tyrosine hydroxylase in human adrenals by a sandwich enzyme immunoassay. *Anal. Biochem. 138*, 125–132.

Nagatsu, T. and Kojima, K. (1985). Application of high-performance liquid chromatography to the study of enzymes and proteins. In K. F. Tipton (Ed.), *Techniques in the Life Sciences, Protein and Enzyme Biochemistry* (in press).

Nagatsu, T., Ichinose, H., Kojima, K., Kameya, T., Shimase, J., Kodama, T., and Shimosato, Y. (1984). Aromatic L-amino acid decarboxylase activities in human lung tissues: Comparison between normal lung and lung carcinomas. *Biochem. Med.* (in press).

Nelson, W. J. and Traub, P. (1982). Purification and further characterization of the Ca^{2+}-activated proteinase specific for the intermediate filament proteins vimentin and desmin. *J. Biol. Chem. 257*, 5544–5553.

Oka, K., Ashiba, G., Sugimoto, T., Matsuura, S., and Nagatsu, T. (1982). Kinetic properties of tyrosine hydroxylase purified from bovine adrenal medulla and bovine caudate nucleus. *Biochim. Biophys. Acta 706*, 188–196.

Parman, A. Ü. and Rideout, J. M. (1983). Purification of porcine proinsulin by high-performance liquid chromatography. *J. Chromatogr. 256*, 283–291.

Rudolph, F. B. and Clark, S. W. (1982). High-performance liquid chromatography of proteins: Purification of the acidic isozyme of adenylosuccinate synthetase from rat liver. *Anal. Biochem. 127*, 193–197.

Togari, A., Kano, H., Oka, K., and Nagatsu, T. (1983). Simultaneous simple purification of tyrosine hydroxylase and dihydropteridine reductase. *Anal. Biochem. 132*, 183–189.

Trocewicz, J., Oka, K., and Nagatsu, T. (1982). Highly sensitive assay for phenylethanolamine N-methyltransferase activity in rat brain by high-performance liquid chromatography with electrochemical detection. *J. Chromatogr. 227*, 407–413.

Progress in HPLC, Vol. 1, pp. 157−177
Parvez *et al.* (Eds)
© 1985 VNU Science Press

High-performance liquid chromatography for enkephalins and enkephalin-containing peptides

KOHICHI KOJIMA and TOSHIHARU NAGATSU

Laboratory of Cell Physiology, Department of Life Chemistry, Graduate School at Nagatsuta, Tokyo Institute of Technology, Yokohama 227, Japan

INTRODUCTION

The endogenous opiate peptides were first discovered and identified as penta-peptides, methionine-enkephalin (Tyr−Gly−Gly−Phe−Met, Met-enkephalin) and leucine-enkephalin (Tyr−Gly−Gly−Phe−Leu, Leu-enkephalin) in 1975 (Hughes *et al.*, 1975), Met-enkephalin was subsequently found to be a part of a previously discovered pituitary peptide, β-lipotropin (β-LPH) with 91 amino acid residues (Li *et al.*, 1965). The sequence from residues 61 to 65 of β-LPH is identical to Met-enkephalin. But no peptide containing the sequence for Leu-enkephalin was known at that time. Within a few years of the discovery of Met- and Leu-enkephalins in pig brain, three enkephalin-containing (EC-) proteins for Met-enkephalin and Leu-enkephalin have been identified as the possible precursors for opioid peptides [pre-proopiomelanocortin (Nakanishi *et al.*, 1979), pre-proenkephalin (pre-pro-enkephalin A) (Noda *et al.*, 1982; Gubler *et al.*, 1982), and pre-prodynorphin (pre-proenkephalin B) (Kakidani *et al.*, 1982)] .

In this chapter, we review the identification and purification of EC-proteins and peptides, structure analysis, and their enzymatic processing by using high-performance liquid chromatography (HPLC).

IDENTIFICATION AND PURIFICATION OF EC-PEPTIDES AND PROTEINS

Soon after the discovery of the two pentapeptides (Met-enkephalin and Leu-enkephalin), precursor EC-proteins were subsequently isolated from the pituitary gland of rat (Rubinstein *et al.*, 1978) and camel (Kimura *et al.*, 1979) using HPLC. Table 1 summarizes isolation and/or purification of EC-peptides and proteins using HPLC.

Before these experiments, smaller precursors, α- and γ-endorphin (Ling *et al.*, 1976), β-lipotropin (Rubinstein *et al.*, 1977a) and β-endorphin (Rubinstein *et al.*, 1977b) were isolated also by HPLC techniques. However, these precursor EC-proteins did not contain a Leu-enkephalin sequence.

Table 1. Isolation and purification of enkephalin-containing peptides and proteins

Starting material	Found or isolated substance(s)	Column(s)	References
Porcine hypothalamus neurohypophysis	α-Endorphin and γ-Endorphin	μBondapak C18	Ling *et al.* (1976)
Rat anterior pituitary	β-Lipotropin	Partisil SCX and LiChrosorb RP-18	Rubinstein *et al.* (1977a)
Rat pituitary	β-Endorphin	Partisil SCX and LiChrosorb RP-18	Rubinstein *et al.* (1977b)
Rat anterior pituitary	Pro-opiocortin	LiChrosorb PR-18	Rubinstein *et al.* (1978)
Camel pituitary	Pro-opiocortin	LiChrosorb RP-8	Kimura *et al.* (1979)
Bovine adrenal medulla and adrenal medullary chromaffin granules	Putative enkephalin precursors (mol. wt 20 000, 10 000, and 5000)	LiChrosorb RP-18	Lewis *et al.* (1979)
Porcine pituitary	Dynorphin-(1–13)	C_{18} column	Goldstein *et al.* (1979)
Porcine hypothalami	Pro-methionine-enkephalin (H–Tyr–Gly–Gly–Phe–Met(O)–Arg–OH)	μBondapak C18	Huang *et al.* (1979)
Bovine adrenal chromaffin granules	Met-enkephalin[Arg[6], Phe[7]] (Tyr–Gly–Gly–Phe–Met–Arg–Phe) and a few other peptides	LiChrosorb RP-18	Stern *et al.* (1979)
Bovine adrenal chromaffin granules	Peptides of 3000–5000 (one has two copies of Met-enkephalin sequence the other has one Met-enkephalin and one Leu-enkephalin)	LiChrosorb RP-18, Ultrasphere Octyl and Spherisorb CN	Kimura *et al.* (1980)
Bovine adrenal chromaffin granules	Proteins of 8000 and 14 000	LiChrosorb RP-18, Ultrasphere ODS, Spherisorb CN and diphenyl-RP	Lewis *et al.* (1980c)
Human adrenal medullary tumour	Met-enkephalin precursors (mol. wt ~2000)	Spherisorb ODS and Partisil ODS	Clement-Jones *et al.* (1980)
Bovine adrenal chromaffin granules, bovine striata and human putamen	Hexa- and Hepta- peptides (Met-enkephalin-Arg[6]–Arg[7], Met-enkephalin-Arg[6]–Phe[7], Met-enkephalin-Lys[6], Met-enkephalin-Arg[6] and Leu-enkephalin-Arg[6])	LiChrosorb RP-18, Ultrasphere ODS and diphenyl-RP	Stern *et al.* (1980)

Table 1 (continued)

Starting material	Found or isolated substance(s)	Column(s)	References
Bovine adrenal medulla	Dodecapeptide (BAM-12P)	μBondapak C-18	Mizuno *et al.* (1980a)
Bovine adrenal medulla	Docosa- and Eicosa-peptide (BAM-22P and BAM-20P)	μBondapak C-18	Mizuno *et al.* (1980b)
Bovine adrenal chromaffin granules	3600 and 4900 enkephalin-containing polypeptides	LiChrosorb RP-18, Ultrasphere ODS and diphenyl-RP	Stern *et al.* (1981)
Bovine adrenal chromaffin granules	3 200 enkephalin-containing peptide (contains both a Met- and Leu-enkephalin sequence)	LiChrosorb RP-18, Ultrasphere ODS and diphenyl-RP	Kilpatrick *et al.* (1981a)
Porcine hypothalami	β-Neo-endorphin	μBondapak C-18	Minamino *et al.* (1981)
Porcine pituitary	Dynorphin-(1–17)	Reverse-phase column	Goldstein *et al.* (1981)
Bovine adrenal medulla chromaffin granules	Met-enkephalin-Arg^6–Gly^7–Leu^8	LiChrosorb RP-18 and Ultrasphere ODS	Kilpatrick *et al.* (1981b)
Bovine adrenal medulla chromaffin granules	A 5300 enkephalin-containing polypeptide (contains single Met-enkephalin-Arg^6–Gly^7–Leu^8 sequence)	LiChrosorb RP-18 and Zorbax TMS	Jones *et al.* (1982a)
Porcine duodenum	Dynorphin-(1–17)	Nucleosil C18 and Nucleosil Phenyl	Tachibana *et al.* (1982)
Bovine adrenal medulla chromaffin granules	18 200 enkephalin-containing polypeptide (contains four Met-enkephalin sequences)	LiChrosorb RP-8, LiChrosorb CN and diphenyl-RP	Kilpatrick *et al.* (1982a)
Bovine posterior pituitary glands	Rimorphin (contains one Leu-enkephalin sequence)	Ultrasphere ODS and Nucleosil phenyl	Kilpatrick *et al.* (1982b)
Porcine pituitary	A 4000 dynorphin	μBondapak C18	Fischli *et al.* (1982)
Bovine posterior pituitary glands	Rimorphin, dynorphin and α-neo-endorphin	Ultrasphere ODS Nucleosil phenyl	Kilpatrick *et al.* (1982c)
Bufo marinus brain	Enkephalins and proenkephalin-derived enkephalin-containing peptides	Nucleosil C18 and Nucleosil phenyl	Kilpatrick *et al.* (1983a)
Bovine caudate nucleus	Tyr–Gly–Gly–Phe–Met–Arg–Arg–Val–NH$_2$	Ultrasphere ODS	Weber *et al.* (1983)

Figure 1 shows the schematic structure and primary amino acid sequences of pre-proopiomelanocortin-related compounds.

In 1979, α-neo-endorphin, a Leu-enkephalin containing peptide, was isolated from porcine hypothalami using a series of conventional column chromatography (Kangawa *et al.*, 1979). This was the first report of the precursor of Leu-enkephalin. α-Neo-endorphin showed very potent opiate activity in guinea pig ileum assay, 6.7 times as high as Met-enkephalin, 5 times as high as β-endorphin and 30 times as high as Leu-enkephalin. Another Leu-enkephalin containing peptide, dynorphin-(1–13), was found in porcine pituitary (Goldstein *et al.*, 1979). It contains Leu-enkephalin sequence at the N-terminal position, but different amino acid sequence at the C-terminal side. These two peptides and related precursor peptides are summarized in Fig. 2. In the guinea pig ileum longitudinal muscle preparation,

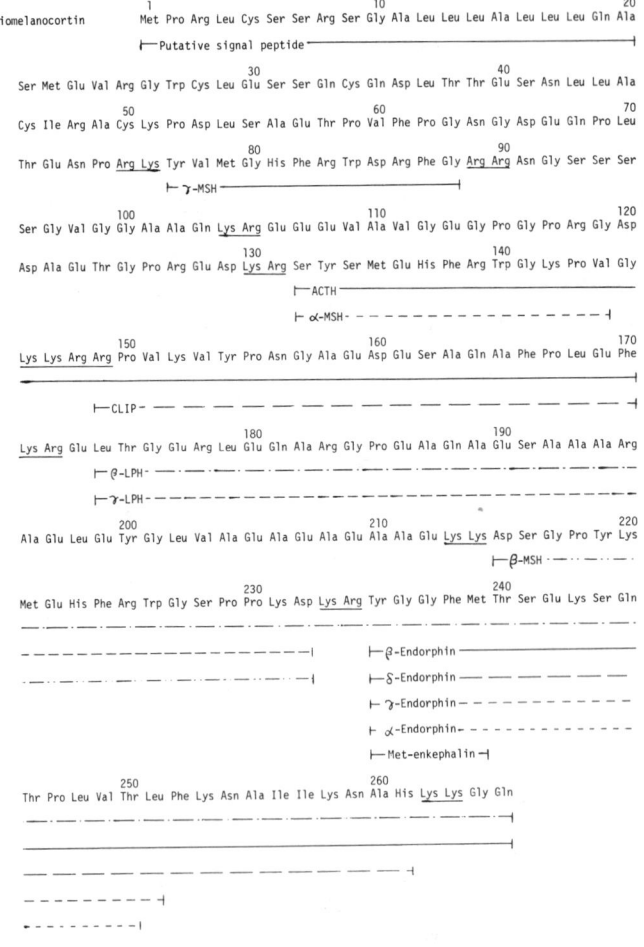

Figure 1. Primary structures of pre-proopiomelanocortin (corticotropin-β-lipotropin precursor) and related polypeptides. The figure is based on the data from Nakanishi *et al.* (1979), Ling *et al.* (1976), and Li *et al.* (1965).

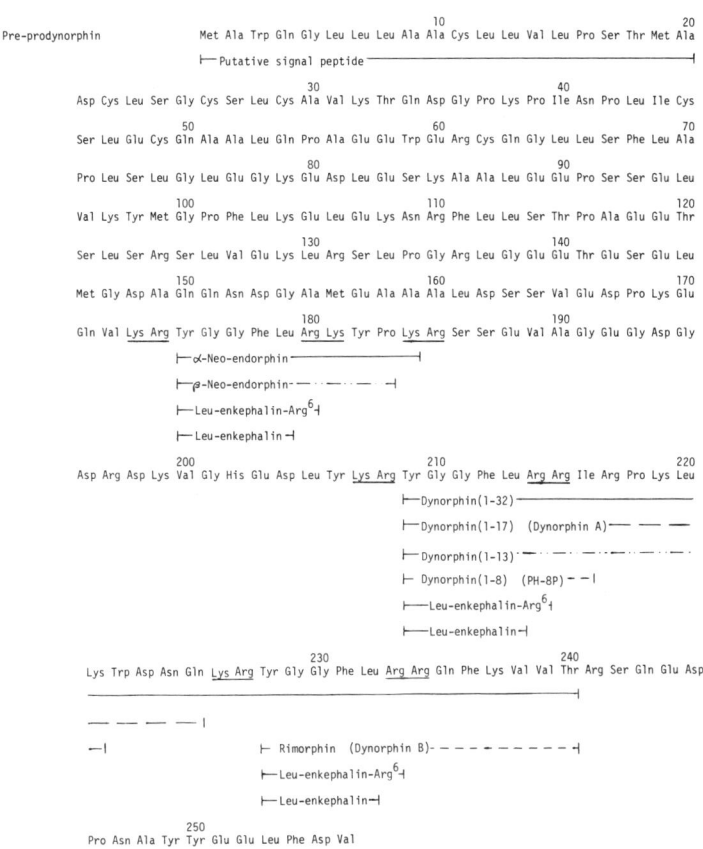

Figure 2. Primary structures of pre-prodynorphin (β-neo-endorphin/dynorphin precursor) and related enkephalin-containing polypeptides. The figure is based on the data from Kakidani *et al.* (1982), Kilpatrick *et al.* (1982b, c), Fischli *et al.* (1982), Tachibana *et al.* (1982), Minamino *et al.* (1981), Goldstein *et al.* (1979, 1981), and Kangawa *et al.* (1979, 1981).

dynorphin is about 700 times more potent than Leu-enkephalin. In the mouse was deferens, it is 3 times more potent than Leu-enkephalin. In bovine adrenals, Lewis *et al.* (1979) found three EC-proteins with approximate molecular weights of 20 000, 10 000, and 5000 which yield peptides with opioid activity by tryptic digestion. Those peptides are chromatographically distinct from the tryptic opioid nonapeptide β-LPH (61–69). Before this report, they identified two large proteins (>40 000 and >100 000) that on treatment with trypsin yield opioid peptides in striatal extracts of guinea pig, rat and cattle using conventional chromatography (Lewis *et al.*, 1978). These new findings support the hypothesis that the enkephalin biosynthetic pathway is distinct from that leading to β-endorphin. 'Pro-Met-enkephalin' peptide (H–Tyr–Gly–Gly–Phe–Met(O)–Arg–OH) was found in the acid extracts of porcine hypothalami (Huang *et al.*, 1979). This peptide is not a fragment of either porcine β-lipotropin or β-endorphin, which suggests that it could be a precursor of Met-enkephalin by a pathway differing from the one usually postulated.

The opioid heptapeptide, Met-enkephalin[Arg[6],Phe[7]] (Tyr−Gly−Gly−Phe− Met−Arg−Phe), was isolated from bovine adrenal medullary granules and striatum with Met-enkephalin, Leu-enkephalin, and a few other small opioid peptides (Stern *et al.*, 1979). This peptide has an analgesic (antinociceptive) activity when administered directly into the cerebral ventricles of mice (Inturrisi *et al.*, 1980). In addition, on a molar basis, Met-enkephalin-Arg[6]−Phe[7], with a median effective dose (ED_{50}) of 38.5 nmol/mouse, is 8 times more potent than Met-enkephalin. This suggests that Met-enkephalin-Arg[6]−Phe[7] may be at least as important as the enkephalins in the postulated enkephalin system mediating pain and analgesia.

Larger EC-peptides were isolated one after another. Those molecular weights are about 3000, 5000 (Kimura *et al.*, 1980), 8000 and 14 000 (Lewis *et al.*, 1980c). Hexa-, hepta-, and larger peptides are summarized in Fig. 3 with related compounds. Peptide F (mol. wt about 3000) contains two copies of the Met-enkephalin, and peptide I (about 5000) contains one each of Met-enkephalin and Leu-enkephalin. The structure of these peptides is also discussed in the next section. The 8000 peptide produced one equivalent of Met-enkephalin, the 14 000 peptide yielded one equivalent of Met-enkephalin and two equivalents of Met-enkephalin-Arg[6]. Two groups of putative human Met-enkephalin precursors (mol. wt *ca.* 2000) from an adrenal medullary tumor were isolated (Clement-Jones *et al.*, 1980). Opioid hexapeptides and heptapeptides have been characterized in the extracts of bovine adrenal, brain and human brain (Stern *et al.*, 1980).

Dodeca- (BAM-12P) (Mizuno *et al.*, 1980a), eicosa- (BAM-20P), and docosa-peptide (BAM-22P) (Mizuno *et al.*, 1980b) have been purified to determine the sequences from the bovine adrenal medulla. In the guinea pig ileum assay, the potency of BAM-22P is 26 times that of Met-enkephalin and 22 times that of human β-endorphin. The activity of BAM-20P seems to be slightly weaker than that of BAM-22P, but its potency is 15 times that of Met-enkephalin, while potency of BAM-12P is only 2 times that of Met-enkephalin. EC-polypeptides with molecular weights of 3600 (peptide B), 4900 (peptide I) (Stern *et al.*, 1981) and 3200 (peptide E) (Kilpatrick *et al.*, 1981a) have been isolated and the structures have been completely determined. The 3600 polypeptide contains Met-enkephain-Arg[6]−Phe[7]. Peptide E is a partial sequence of peptide I. Peptide E was 30 times more potent

```
                                                  10                                        20
Pre-proenkephalin              Met Ala Arg Phe Leu Gly Leu Cys Thr Trp Leu Leu Ala Leu Gly Pro Gly Leu Leu Ala
                               |—Putative signal peptide ——————————————————————————————————————

                                                  30                              40
            Thr Val Arg Ala Glu Cys Ser Gln Asp Cys Ala Thr Cys Ser Tyr Arg Leu Ala Arg Pro Thr Asp Leu Asn Pro
            —————————|
                      |—18.2kDal EC-polypeptide —————————————————————————————————————————————
                      |—12.6kDal EC-polypeptide — —  —  —  —  —  —  —  —  —  —  —  —  —  —  —  —
                      |—8.6kDal EC-polypeptide— · — ·· — — — —· — ·— —· —· — ·— —· — —· —· — ·—
             50                                    60                                70
            Leu Ala Cys Thr Leu Glu Cys Glu Gly Lys Leu Pro Ser Leu Lys Thr Trp Glu Thr Cys Lys Glu Leu Leu Gln

            — —  — —  — —  — —  — —  — —  — —  — —  — —  — —  — —  — —  — —  —  — —·
            —· — · — — · — · — · — —  — · — — ·— — · — — —· — · — — · — · — —· — ——·
```

Figure 3. (Continued on page 163.)

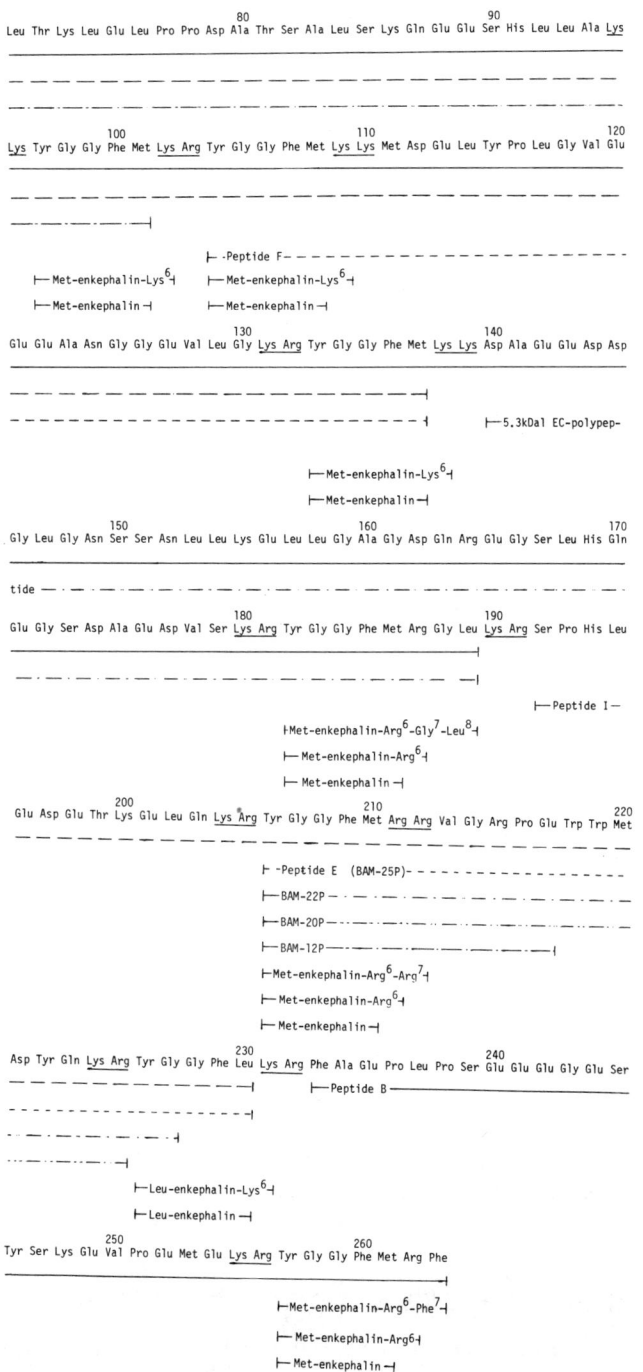

Figure 3. Primary structures of pre-proenkephalin and related enkephalin-containing poly-peptides. The figure is based on the data from Noda *et al*. (1982), Gubler *et al*. (1982), Kilpatrick *et al*. (1981a, b, 1982a), Jones *et al*. (1980b, 1982a, b), Mizuno *et al*. (1980a, b), Stern *et al*. (1979, 1980), and Huang *et al*. (1979).

than Met-enkephalin in the guinea pig ileum assay, which suggests that the adrenal EC-peptides may perform a unique biological function *in vivo*.

From chromaffin granules of bovine adrenal medulla, opioid octapeptide (Tyr– Gly–Gly–Phe–Met–Arg–Gly–Leu, Met-enkephalin-Arg[6]–Gly[7]–Leu[8]) (Kilpatrick *et al.*, 1981b) and a octapeptide-containing polypeptide (mol. wt 5300) (Jones *et al.*, 1982a) was also isolated.

The structure of the complete proenkephalin protein was unknown, while numerous adrenal EC-peptides and EC-peptides from other tissues were isolated. Early in 1982 the complete sequence of bovine proenkephalin was described simultaneously by two groups (Noda *et al.*, 1982; Gubler *et al.*, 1982). Both results suggest that proenkephalin has 30 000 molecular weight. An 18 200 adrenal EC-polypeptide was purified and shown to contain four internal Met-enkephalin sequence (Kilpatrick *et al.*, 1982a). This protein is the largest EC-polypeptide yet isolated from animal tissues. A proenkephalin-like precursor was identified in acid extracts of the brain of *Bufo marinus* (Kilpatrick *et al.*, 1983a). An amidated opioid octapeptide (Tyr–Gly–Gly–Phe–Met–Arg–Arg–Val–NH$_2$) was found from bovine brain which derived by proteolytic cleavage from proenkephalin (Weber *et al.*, 1983). This peptide, named 'metorphamide', was found to have a high opioid μ-receptor-binding activity.

Two types of Leu-enkephalin containing peptides was isolated. One was named β-neo-endorphin (Minamino *et al.*, 1981), the other was named dynorphin-(1–17) (Goldstein *et al.*, 1981).

Elucidation of the structures of proopiomelanocortin and proenkephalin still did not account for the two other Leu-enkephalin-containing peptides, dynorphin and α-neo-endorphin. Dynorphin-(1–17) from porcine duodenum (Tachibana *et al.*, 1982) and rimorphin from bovine posterior pituitary glands (Kilpatrick *et al.*, 1982b) were identified. Rimorphin was the third discovery of Leu-enkephalin containing peptides. Kilpatrick *et al.* (1982c) have reported that rimorphin was found in association with dynorphin and α-neo-endorphin in the bovine posterior pituitary glands. At the same time, Fischli *et al.* (1982) isolated a peptide of 32 amino acid residues from pig pituitary glands that contains dynorphin-(1–17) at its amino terminus and rimorphin sequence at its carboxyl terminus. These finding suggest the existence of a large Leu-enkephalin-containing protein. Sequencing of the cloned cDNA from pig hypothalamus showed a precursor that contains one sequence each of α-neo-endorphin, dynorphin and rimorphin (Kakidani *et al.*, 1982). Researches on EC-peptides described here have been successively carried out by HPLC. A reverse-phase column, mainly C$_{18}$ column, has been used in most cases (Table 1). Udenfriend's group has examined other types of column supports and eluting conditions (Jones *et al.*, 1980a; Lewis *et al.*, 1980a).

STRUCTURE AND SEQUENCE ANALYSIS FOR EC-PEPTIDES BY HPLC

In this section, we describe the method of structure and sequence analysis of EC-peptides by HPLC (Table 2). Lewis *et al.* (1978) reported the presence of high molecular weight proteins in the striatum that can be fragmented with trypsin to yield peptides with opioid activity. The trypsin digests were resolved by a reverse-phase column with an automated preparative/analytical fluorescent peptide analyzer which

Table 2. Structure studies and sequence analyses of enkephalin-containing peptides and proteins by HPLC

Material	Column(s)	References
High molecular weight opioid-containing proteins (>40 000 and >100 000)	LiChrosorb RP-18	Lewis *et al.* (1978)
Dynorpin-(1–13)	Zorbax CN	Goldstein *et al.* (1979)
A common precursor of Met- and Leu-enkephalin (*ca.* 50 000)	LiChrosorb RP-18	Lewis *et al.* (1980b)
Enkephalin-containing polypeptides (4000 and 5000)	Zorbax ODS and Zorbax CN	Jones *et al.* (1980b)
Enkephalin-containing polypeptides (3600 and 4900)	LiChosorb RP-18, Zorbax ODS and Zorbax CN	Stern *et al.* (1981)
α-Neo-endorphin	μBondapak C-18	Kangawa *et al.* (1981)
Met-enkephalin-Arg[6]–Gly[7]–Leu[8]	LiChrosorb RP-18	Kilpatrick *et al.* (1981b)
Dynorphin-(1–17)	Nucleosil C18	Tachibana *et al.* (1982)
Enkephalin-containing proteins (8600 and 12 600)	Zorbax TMS, Ultraspher ODS, Zorbax ODS and Zorbax CN	Jones *et al.* (1982b)
α-Endorphin and γ-endorphin	CP-Spher C8	Burbach and Wiegant (1984)

they developed (Böhlen *et al.*, 1975). They improved the method to get better resolution for Met-enkephalin, Leu-enkephalin, (O)-Met-enkephalin, Met-enkephalin-Arg[6], Met-enkephalin-Lys[6], Leu-enkephalin-Arg[6], Met-enkephalin-Arg[6]–Arg[7], Met-enkephalin-Arg[6]–Phe[7] (Stern *et al.*, 1980) and then Met-enkephalin-Arg[6]–Gly[7]–Leu[8] (Kilpatrick *et al.*, 1981b). Using this method, Lewis *et al.* (1980b) showed the ratio of Met-enkephalin to Leu-enkephalin in the precursors. After purification of EC-peptides, the content of Met-enkephalin and Leu-enkephalin were determined by this method (Jones *et al.*, 1980b, 1982b; Udenfriend *et al.*, 1983). The amino acid composition and amino terminus analysis have been done by HPLC based on dansylation, Edman degradation and derivatization with phenylthiohydantoin or *o*-phthaldialdehyde. Udenfriend's group effectively applied HPLC to the picomole level detection of EC-peptides using an amino acid analyzer with fluorescamine fluorometry (Stein *et al.*, 1973).

EXAMINATION OF TISSUE DISTRIBUTION OF ENKEPHALIN-RELATED SUBSTANCES BY HPLC WITH IMMUNOCHEMICAL DETECTION

In addition to identification and/or purification studies, the tissue distribution of enkephalins and EC-peptides has been extensively studied. Radioimmunological and immunohistological methods have been used to study the distribution of

Met-enkephalin, Leu-enkephalin and other EC-peptides. But these methods are not suitable to quantify the amount of EC-peptides based on the cross-reactivity of the antisera. However, the method of combination of radioimmunoassay and HPLC has made it possible to determine the exact amount of EC-peptides after they are separated using a calibrated HPLC column. The specificity of this HPLC-immunochemical detection is like a two dimensional electrophoresis or TLC.

In this section, we describe the tissue distribution of EC-peptides of different species by immunological assay with HPLC or by radioreceptor and/or radiobinding assay with HPLC (Table 3).

Met-enkephalin-Arg^6–Phe^7 is a good marker for proenkephalin and has interesting biological activities. Distribution of this heptapeptide followed that of Met-enkephalin in rat brain, with the highest concentrations in the globus pallidus, intermediate levels in caudate-putamen and hypothalamus, and low levels in cortex and cerebellum (Boarder et al., 1982a). Boarder et al. (1982b) have also compared bovine caudate and adrenal using the heptapeptide antiserum, and they have suggested different processing of the peptide in both tissues. However, Ikeda et al. (1982) have measured heptapeptide with Met-enkephalin, Leu-enkephalin and Met-enkephalin-Arg^6–Gly^7–Leu^8 simultaneously using HPLC and radioimmunoassays in the brain of guinea pig, rat, and golden hamster. They have demonstrated the co-existence of the four peptides at a constant ratio which proenkephalin had and also suggested that biosynthetic pathway in the brain is similar to that in the adrenal medulla. Tang et al. (1982) have measured heptapeptide in extracts of various peripheral organs, and found the highest concentration in ileum and lung. They have also suggested that Met-enkephalin-Arg^6–Phe^7 may not only be a precursor of Met-enkephalin but may also act as a neurotransmitter and neuromodulator by itself. They (Tang et al., 1983) have also demonstrated that heptapeptide might be an important endogenous opiate peptide located in APUD (Amine Precursor Uptake and Decarboxylation) cells of the lung of several species. They have also reported heptapeptide content of human and rabbit serum and postulated its physiological role in respiratory or cardiocirculatory function (Chou et al., 1983).

On the other hand, the distribution of prodynorphin-related peptides is more complex. For example, Suda et al. (1982) have examined rat anterior pituitary (AP), intermediate-posterior pituitary (IP), and medial basal hypothalamus (MBH), and have suggested (1) the presence of dynorphin-(1–17), -(1–13), -(1–11), and -(1–10) in IP, (2) dynorphin-(1–11) and -(1–10) as the major components, and (3) the difference of processing in AP, IP, and MBH. Two studies with rat brain have been reported using antisera to α-neo-endorphin and β-neo-endorphin with HPLC (Weber et al., 1982; Kitamura et al., 1982). Weber et al. (1982) showed that in hypothalamus and posterior pituitary β-neo-endorphin levels were almost as high as α-neo-endorphin levels. However, in the striatum α-neo-endorphin was 30-fold more concentrated than β-neo-endorphin and in all other brain regions α-neo-endorphin was present in 3- to 20-fold higher concentrations than β-neo-endorphin. These results showed that β-neo-endorphin was concentrated in the pituitary, especially in the posterior lobe, like α-neo-endorphin. β-Neo-endorphin was also widely distributed in brain, but in a different manner from α-neo-endorphin.

Table 3. Tissue distribution of enkephalin-containing peptides examined by HPLC

Material	Tissue(s)	Column(s)	References
Met-enkephalin-Arg[6]–Phe[7]	Rat brain	Altex C-18 ultrasphere-ODS	Boarder *et al.* (1982a)
Met-enkephalin-Arg[6]–Phe[7] and Met-enkephalin	Bovine caudate and bovine medulla chromaffin granule	Altex C-18 ultrasphere-ODS	Boarder *et al.* (1982b)
Enkephalin-containing polypeptides	Guinea pig adrenal, brain, and myenteric plexus	LiChrosorb RP-18 and Hypersil ODS	Kojima *et al.* (1982)
Multiple forms of immunoreactive dynorphin	Rat anterior pituitary, intermediate-posterior pituitary, and medial basal hypothalamus	Nucleosil C18	Suda *et al.* (1982)
Met-enkephalin-Arg[6]–Phe[7]	Peripheral organs of rat and guinea pig	Altex C-18 ultrasphere ODS-5	Tang *et al.* (1982)
Met-enkephalin-Arg[6]–Gly[7]– Leu[8], Met-enkephalin, Leu-enkephalin, and Met-enkephalin-Arg[6]–Phe[7]	Brain of guinea pig, rat and golden hamster	Ultrasphere ODS	Ikeda *et al.* (1982)
α-Neo-endorphin and β-neo-endorphin	Rat brain	Reverse phase column	Weber *et al.* (1982)
α-Neo-endorphin and β-neo-endorphin	Rat brain and pituitary	LS-410 ODS SIL	Kitamura *et al.* (1982)
Met-enkephalin-Arg[6]–Phe[7], Met-enkephalin, Leu-enkephalin, dynorphin-(1–13) and α-neo-endorphin	Rat brain	Ultrasphere C8	Giraud *et al.* (1983)
Met-enkephalin-Arg[6]–Phe[7]	Rat, guinea pig, and human lung	Altex C18 ultrasphere-ODS	Tang *et al.* (1983)
Met-enkephalin-Arg[6]–Phe[7]	Human and rabbit plasma	Altex C18 ultrasphere-ODS	Chou *et al.* (1983)
Dynorphin, α-neo-endorphin, Met-enkephalin, and substance P	Human spinal cord and spinal ganglia	μBondapak C18	Przewtocki *et al.* (1983)
Met-enkephalin and Leu-enkephalin	Guinea pig hippocampus	μBondapak C18	Hoffman *et al.* (1983)
Dynorphin-(1–17), β-neo-endorphin, Leu-enkephalin, rimorphin, and dynorphin-(1–24)	Porcine neurointermediate lobe	Nucleosil C18 and Nucleosil phenyl	Kilpatrick *et al.* (1983b)
Met-enkephalin and Leu-enkephalin	Guinea pig retina	μBondapak C18	Hoffman (1983)
Dynorphin-(1–17), dynorphin B, dynorphin-(1–8), α-neo-endorphin, and β-neo-endorphin	Rat brain and pituitary	μBondapak C18	Seizinger *et al.* (1984)

Remarkable differences were observed in hippocampus, striatum and cortex, where α-neo-endorphin was found to predominate over β-neo-endorphin. These results suggest that in certain brain regions such processing mechanisms exist that can generate β-neo-endorphin through processing of α-neo-endorphin or its precursors.

EC-peptides related both to proenkephalin and to prodynorphin were studied simultaneously (Giraud et al., 1983) and distinct tissue distribution of both precursors was found. Prodynorphin related EC-peptides were examined simultaneouly as proenkephalin related peptides (Kilpatrick et al., 1983b; Seizinger et al., 1984). Kilpatrick et al. have quantitated the EC-peptides in porcine pituitary neurointermediate lobe. Dynorphin-(1–17), β-neo-endorphin, Leu-enkephalin, rimorphin, and dynorphin were found in the highest concentrations (200–700 pmol/g), but no detectable amount of rimorphin-(1–29) was found. Seizinger et al. (1984) have investigated the distribution of EC-peptide in rat brain and pituitary. Their results also support the possibility that differential proteolytic processing of prodynorphin occurs within different regions of brain and pituitary. Moreover, evidence is provided that, in addition to the paired basic amino acids −Lys−Arg− as the typical cleavage site for peptide hormone precursors, other cleavage signals also seem to exist for the processing of prodynorphin and proenkephalin.

ANALYSIS OF ENZYMES FOR PROCESSING OF ENKEPHALIN AND EC-PEPTIDES BY HPLC

The enkephalin-degrading enzymes have already long history. Since many EC-peptides related to enkephalins have been identified and purified from several tissues and species, the search for processing enzymes of enkephalins and EC-peptides has started. In this section, we describe studies on enkephalin-processing enzyme with HPLC. For enkephalin-degrading enzymes, some reviews are listed in the References (Schwartz et al., 1981; Hersh, 1982).

Table 4 shows substrates, enzyme(s), products and HPLC column. Small peptide substrates are mainly used for this purpose. Hexapeptide (Met-enkephalin-Arg^6 or Lys^6, and Leu-enkephalin-Arg^6 or Lys^6) and heptapeptide (Met-enkephalin-Arg^6 − Arg^7 or -Arg^6−Lys^7 or -Arg^6−Phe^7 and Leu-enkephalin-Arg^6−Arg^7) have been digested by enkephalin convertase (carboxypeptidase) (Fricker and Snyder, 1982) or by dipeptidyl carboxypeptidase (Demmer and Braud, 1983) to yield Met-enkephalin and Leu-enkephalin. Dipeptidyl carboxypeptidase also yielded Tyr−Gly−Gly−Phe. Rat brain or kidney metalloendopeptidase generated Tyr−Gly−Gly and Phe−Met−Arg−Phe from heptapeptide (Benuck et al., 1982). The latter peptide has a cardioactive properties. Orlowski and Wilk (1981) also used heptapeptide (Leu-enkephalin-Arg^6−Phe^7−2NA) and found a multicatalytical neutral endopeptidase complex which generates Leu-enkephalin-Arg^6 (16%), Leu-enkephalin (30%) and Leu-enkephalin-Arg^6−Phe^7 (54%). A proenkephalin processing enzyme from bovine adrenal has cleaved BAM-12P to yield Met-enkephalin-Arg^6 and Met-enkephalin (Evangelista et al., 1982). Endogenous precursor(s) has also been used as substrate. An enkephalin-generating enzyme has produced Met-enkephalin and Arg^1-Met-enkephalin (Lindberg et al., 1982). In the chromaffin granule lysate, Troy and Musacchio (1982) have found enzymes which cleaved endogenous precursors

Table 4. Enzymes for processing of enkephalin-containing peptides

Substrate(s)	Enzyme(s)	Product(s)	Column	References
Leu-enkephalin-Arg^6–Phe^7–2NA	A multicatalytical neutralendopeptidase complex (~700 000, bovine pituitary)	2NA, Leu-enkephalin-Arg^6 (16%) Leu-enkephalin (30%), Leu-enkephalin-Arg^6–Phe^7 (54%), and Phe–2NA	μBondapak C18	Orlowski and Wilk (1981)
Met-enkephalin-Arg^6, Leu-enkephalin-Arg^6, Met-enkephalin-Lys^6, and Leu-enkephalin-Lys^6	Enkephalin convertase (carboxy peptidase, stimulated by CoCl$_2$, inhibited by EDTA or 1,10-phenanthroline)	Met-enkephalin and Leu-enkephalin	μBondapak C18	Fricker and Snyder (1982)
Endogenous substrate(s)	An enkephalin-generating enzyme (a serine protease, bovine adrenal medulla)	Met-enkephalin and Arg^1-Met-enkephalin	BioSil ODS-10	Lindberg et al. (1982)
BAM-12 and Leu-8	A proenkephalin processing enzyme (a trypsin-like enzyme, optimum pH 5, bovine adrenal chromaffin granule)	Met-enkephalin-Arg^6 and Met-enkephalin	AllTech C18	Evangelista et al. (1982)
Endogenous precursor(s) and Peptide E	Chromaffin granule enzymes (bovine adrenal)	Met-enkephalin, Leu-enkephalin, and Met-enkephalin-Arg^6	μBondapak C18	Troy and Musacchio (1982)
Tyr–Gly–Gly–Phe–Met–Arg–Phe	Metalloendopeptidase (rat brain and kidney)	Tyr–Gly–Gly and Phe–Met–Arg–Phe (cardioactive neuropeptide)	Spherisob 5μ ODS	Benuck et al. (1982)

Table 4 (continued)

Substrate(s)	Enzyme(s)	Product(s)	Column	References
Endogenous proenkephalin and BAM-12P	Proenkephalin-converting enzyme (bovine adrenal chromaffin granules, mol. wt ca. 220 000, thiol dependent protease, optimal pH 5.5)	Met-enkephalin	μBondapak C18	Mizuno et al. (1982)
Met-enkephalin-Arg^6, Met-enkephalin-Arg^6–Arg^7, Met-enkephalin-Arg^6–Lys^7, Met-enkephalin-Arg^6–Phe^7, and Leu-enkephalin-Arg^6–Arg^7	Dipeptidyl carboxypeptidase (rat brain cortical synaptic plasma membrane, inhibited by metal chelating agents and thiols but not by inhibitors of serine proteases, thermolysin and enkephalinase)	Tyr–Gly–Gly–Phe, Met-enkephalin, and Leu-enkephalin	LiChrosorb RP-18	Demmer and Brand (1983)
Endogenous substrate(s)	Bovine adrenal medulla chromaffin granules	Enkephalin-containing peptides	LiChrosorb RP-18	Fleminger et al. (1983)

and peptide F to yield Met-enkephalin, Leu-enkephalin and Met-enkephalin-Arg[6]. Mizuno *et al.* (1982) have used endogenous precursor(s) and BAM-12P as substrate, and found Met-enkephalin as product. Fleminger *et al.* (1983) have used endogenous substrate and found a gradual shift toward lower molecular weight EC-peptides during the incubation (37°C for up to 22 h), indicating processing of the EC-peptides from the higher molecular weight forms to the lower molecular weight forms. The total amount of Met-enkephalin remained constant during incubation. HPLC resolution of the fraction containing the low molecular weight EC-peptides showed that free enkephalins as well as Met-enkephalin-Arg[6]–Phe[7] and Met-enkephalin-Arg[6]–Gly[7]–Leu[8] were accumulated while Met-enkephalin-Arg[6] and Met-enkephalin-Lys[6] disappeared.

All the above data indicate the presence of an atypical trypsin-like activity and the presence of a carboxypeptidase B-like activity within the chromaffin granules.

OTHER RELATED STUDIES ON ENKEPHALINS USING HPLC

Table 5 shows other related studies on enkephalins using HPLC. Several studies have been carried out using cultured cells. Rossier *et al.* (1980) showed that [^{35}S]methionine is incorporated into the Met-enkephalin sequences of a 22 000 molecular weight EC-protein in the adrenal medulla and further indicated the processing to Met-enkephalin and Met-enkephalin-Arg[6]–Phe[7]. Livett *et al.* (1981) reported that Leu-enkephalin and catecholamines (total catecholamines or nor-adrenaline and adrenaline) are released together from primary cultures of bovine adrenal medullary chromaffin cells by nicotine in a Ca^{2+}-dependent manner. Rossier *et al.* (1981) also reported the levels of enkephalin congeners and their precursors.

Dandekar and Sabol (1982a, b) studied the biosynthesis of the protein precursor of Met-enkephalin using cell-free translation systems to characterize the enkephalin-precursor gene product and mRNA of bovine adrenals (1982a) and bovine striatum (1982b). They also suggested that a protein similar or identical to bovine adrenal medullary pre-proenkephalin is the major Met-enkephalin precursor synthesized in brain as well as adrenal medulla, and that pre-proenkephalin is converted to a protein resembling proenkephalin, presumably by removal of a signal peptide. Tanaka *et al.* (1982) reported the expression in *Escherichia coli* of chemically synthesized gene for a novel opiate peptide α-neo-endorphin.

Kilpatrick *et al.* (1980) demonstrated that each of the EC-peptides and -proteins stored in chromaffin vesicles was secreted from perfused bovine adrenal glands after stimulation with nicotine or Ba^{2+}, the agents that release catecholamines.

Possenti *et al.* (1983) examined enkephalin-binding systems in human plasma by HPLC with the combination of reverse-phase and gel-filtration columns.

CONCLUSIONS

Since the discovery of Met-enkephalin and Leu-enkephalin in 1975, opioid peptides and its precursors have been effectively studied using HPLC. The data on the primary structures of EC-peptides shows great similarities between human and bovine proenkephalin. Several structural similarities also exist among the three precursors, pre-proopiomelanocortin, pre-proenkephalin and pre-prodynorphin,

Table 5. Various studies on enkephalin-containing peptides by HPLC

Material(s)	Experiment(s)	Column(s)	References
22 000 Met-enkephalin-containing protein and enkephalins	[^{35}S] Methionine incorporation and pulse-chase experiments	Ultrasphere RP-8	Rossier *et al.* (1980)
Enkephalins and enkephalin-containing polypeptides	Perfusion of bovine adrenal gland	LiChrosorb RP-18	Kilpatrick *et al.* (1980)
Leu-enkephalin	Release from primary cultures of bovine adrenal medullary chromaffin cells	Ultrasphere RP-8	Livett *et al.* (1981)
Enkephalins and its precursors	Release from primary cultures of adrenal chromaffin cells	Ultrasphere RP-8	Rossier *et al.* (1981)
Enkephalin-precursor protein	Cell-free translation (bovine adrenal medullary mRNA)	μBondapak C18	Dandekar and Sabol (1982a)
Proenkephalin	Cell-free translation (bovine striatum mRNA)	μBondapak C18	Dandekar and Sabol (1982b)
A novel opiate peptide α-neo-endorphin	Expression of α-neo-endorphin (synthetic)	μBondapak C18	Tanaka *et al.* (1982)
Methionine-enkephalin precursor protein	*In vitro* biosynthesis and initial processing (bovine adrenal medulla and striatum)	μBondapak C18	Sabol *et al.* (1983)
Leu-enkephalin and Met-enkephalin	Enkephalin-binding system (human plasma)	LiChrosorb RP-18 TSK G2000SW and LiChrosorb Diol	Possenti *et al.* (1983)

besides the presence of enkephalin sequences flanked by paired basic residues. These similarities suggest an evolutionary relationship among the three different precursors.

An explanation of the multiple enkephalin copies in a single precursor which many researchers are now favoring is that it is the EC-peptides and not the free enkephalins that are the physiological agents.

Though structures of the three precursors of enkephalins have been revealed, many questions are still remaining especially for the processing, and physiological and pharmacological roles. We have reviewed only the studies on EC-peptide by HPLC. However, many studies have also done without HPLC. We include the references on physiological and pharmacological aspects of EC-peptides (Bloom, 1983; Holaday, 1983; Frederickson and Geary, 1982; Lewis and Stern, 1983; Udenfriend and Kilpatrick, 1983; Hughes, 1983).

REFERENCES

Benuck, M., Berg, M. J., and Marks, N. (1982). Rat brain and kidney metalloendopeptidase: enkephalin heptapeptide conversion to form a cardioactive neuropeptide, Phe–Met–Arg–Phe-amide. *Biochem. Biophys. Res. Commun. 107*, 1123–1129.

Bloom, F. E. (1983). The endorphins: a growing family of pharmacologically pertinent peptides. *Ann. Rev. Pharmacol. Toxicol. 23*, 151–170.

Boarder, M. R., Lockfeld, A. J., and Barchas, J. D. (1982a). Measurement of methionine-enkephalin[Arg⁶, Phe⁷] in rat brain by specific radioimmunoassay directed at methionine sulphoxide enkephalin[Arg⁶, Phe⁷]. *J. Neurochem. 38*, 299–304.

Boarder, M. R., Lockfeld, A. J., and Barchas, J. D. (1982b). Met-enkephalin[Arg⁶, Phe⁷] immunoreactivity in bovine caudate and bovine adrenal medulla. *J. Neurochem. 39*, 149–154.

Böhlen, P., Stein, S., Stone, J., and Udenfriend, S. (1975). Automatic monitoring of primary amines in preparative column effluents with fluorescamine. *Anal. Biochem. 67*, 438–445.

Burbach, J. P. H. and Wiegant, V. M. (1984). Isolation and characterization of α-endorphin and γ-endorphin from single human pituitary glands. *FEBS Lett. 166*, 267–272.

Chou, J., Tang, J., and Costa, E. (1983). Met⁵-enkephalin-Arg⁶–Phe⁷ content of human and rabbit plasma. *Life Sci. 32*, 2589–2595.

Clement-Jones, V., Corder, R., and Lowry, P. J. (1980). Isolation of human met-enkephalin and two groups of putative precursors (2K-pro-met-enkephalin) from an adrenal medullary tumour. *Biochem. Biophys. Res. Commun. 95*, 665–673.

Dandekar, S. and Sabol, S. L. (1982a). Cell-free translation and partial characterization of mRNA coding for enkephalin-precursor protein. *Proc. Natl. Acad. Sci. USA 79*, 1017–1021.

Dandekar, S. and Sabol, S. L. (1982b). Cell-free translation and partial characterization of proenkephalin messenger RNA from bovine striatum. *Biochem. Biophys. Res. Commun. 105*, 67–74.

Demmer, W. and Brand, K. (1983). A dipeptidyl carboxypeptidase in brain synaptic membranes active in the metabolism of enkephalin containing peptides. *Biochem. Biophys. Res. Commun. 114*, 804–812.

Evangelista, R., Ray, P., and Lewis, R. V. (1982). A 'trypsin-like' enzyme in adrenal chromaffin granules: a proenkephalin processing enzyme. *Biochem. Biophys. Res. Commun. 106*, 895–902.

Fischli, W., Goldstein, A., Hunkapiller, M. W., and Hood, L. E. (1982). Isolation and amino acid sequence analysis of a 4000-dalton dynorphin from porcine pituitary. *Proc. Natl. Acad. Sci. USA 79*, 5435–5437.

Fleminger, G., Ezra, E., Kilpatrick, D. L., and Udenfriend, S. (1983). Processing of enkephalin-containing peptides in isolated bovine adrenal chromaffin granules. *Proc. Natl. Acad. Sci. USA 80*, 6418–6421.

Frederickson, R. C. A. and Geary, L. E. (1982). Endogenous opioid peptides: review of physiological, pharmacological and clinical aspects. *Prog. Neurobiol. 19*, 19–69.

Fricker, L. D. and Snyder, S. H. (1982). Enkephalin convertase: purification and characterization of a specific enkephalin-synthesizing carboxypeptidase localized to adrenal chromaffin granules. *Proc. Natl. Acad. Sci. USA 79*, 3886–3890.

Giraud, P., Castanas, E., Patey, G., Oliver, C., and Rossier, J. (1983). Regional distribution of methionine-enkephalin-Arg⁶–Phe⁷ in the rat brain: comparative study with the distribution of other opioid peptides. *J. Neurochem. 41*, 154–160.

Goldstein, A., Tachibana, S., Lowney, L. I., Hunkapiller, M., and L. Hood. (1979). Dynorphin-(1–13), an extraordinarily potent opioid peptide. *Proc. Natl. Acad. Sci. USA 76*, 6666–6670.

Goldstein, A., Fischli, W., Lowney, L. I., Hunkapiller, M., and Hood, L. (1981). Porcine pituitary dynorphin: complete amino acid sequence of the biologically active heptadecapeptide. *Proc. Natl. Acad. Sci. USA 78*, 7219–7223.

Gubler, U., Seeburg, P., Hoffman, B. J., Gage, L. P., and Udenfriend, S. (1982). Molecular cloning establishes proenkephalin as precursor of enkephalin-containing peptides. *Nature* 295, 206–208.

Hersh L. B. (1982). Degradation of enkephalins: the search for an enkephalinase. *Mol. Cell. Biochem.* 47, 35–43.

Hoffman D. W. (1983). Chromatographic identification of enkephalins in the guinea pig retina. *Neurosci. Lett.* 40, 67–73.

Hoffman, D. W., Altschuler, R. A., and Gutierrez, J. (1983). Multiple molecular forms of enkephalins in the guinea pig hippocampus. *J. Neurochem.* 41, 1641–1647.

Holaday, J. W. (1983). Cardiovascular effects of endogenous opiate systems. *Ann. Rev. Pharmacol. Toxicol.* 23, 541–594.

Huang, W.-Y., Chang, R. C. C., Kastin, A. J., Coy, D. H., and Schally, A. V. (1979). Isolation and structure of pro-methionine-enkephalin: potential enkephalin precursor from porcine hypothalamus. *Proc. Natl. Acad. Sci. USA* 76, 6177–6180.

Hughes, J. (1983). Biogenesis, release and inactivation of enkephalins and dynorphins. *Brit. Med. Bull.* 39, 17–24.

Hughes, J., Smith, T. W., Kosterlitz, H. W., Fothergill, L. A., Morgan, B. A., and Morris, H. R. (1975). Identification of two related pentapeptides from the brain with potent opiate agonist activity, *Nature* 258, 577–579.

Ikeda, Y., Nakao, K., Yoshimasa, T., Yanaihara, N., Numa, S., and Imura, H. (1982). Existence of Met-enkephalin-Arg[6]–Gly[7]–Leu[8] with Met-enkephalin, Leu-enkephalin and Met-enkephalin-Arg[6]–Phe[7] in the brain of guinea pig, rat and golden hamster. *Biochem. Biophys. Res. Commun.* 107, 656–662.

Inturrisi, C. E., Umans, J. G., Wolff, D., Stern, A. S., Lewis, R. V., Stein, S., and Udenfriend, S. (1980). Analgesic activity of the naturally occurring heptapeptide [Met] enkephalin-Arg[6]–Phe[7]. *Proc. Natl. Acad. Sci. USA* 77, 5512–5514.

Jones, B. N., Lewis, R. V. Pääbo, S., Kojima, K., Kimura, S., and Stein, S. (1980a). Effects of flow rate and eluant composition on the high performance liquid chromatography of proteins. *J. Liq. Chromatogr.* 3, 1373–1383.

Jones, B. N., Stern, A. S., Lewis, R. V., Kimura, S., Stein, S., Udenfriend, S., and Shively, J. E. (1980b). Structure of two adrenal polypeptides containing multiple enkephalin sequences. *Arch. Biochem. Biophys.* 204, 392–395.

Jones, B. N., Shively, J. E., Kilpatrick D. L., Kojima, K., and Udenfriend, S. (1982a). Enkephalin biosynthetic pathway: a 5300-dalton adrenal polypeptide that terminates at its COOH end with the sequence [Met] enkephalin-Arg–Gly–Leu–COOH. *Proc. Natl. Acad. Sci. USA* 79, 1313–1315.

Jones, B. N., Shively, J. E., Kilpatrick, D. L., Stern, A. S., Lewis, R. V., Kojima, K., and Udenfriend, S. (1982b). Adrenal opioid proteins of 8600 and 12 600 daltons: intermediates in proenkephalin processing. *Proc. Natl. Acad. Sci. USA* 79, 2096–2100.

Kakidani, H., Furutani, Y., Takahashi, H., Noda, M., Morimoto, Y., Hirose, T., Asai, M., Inayama, S., Nakanishi, S., and Numa, S. (1982). Cloning and sequence analysis of cDNA for porcine β-neo-endorphin/dynorphin precursor. *Nature* 298, 245–249.

Kangawa, K., Matsuo, H., and Igarashi, M. (1979). α-Neo-endorphin: A 'big' leu-enkephalin with potent opiate activity from porcine hypothalami. *Biochem. Biophys. Res. Commun.* 86, 153–160.

Kangawa, K., Minamino, N., Chino, N., Sakakibara, S., and Matsuo, H. (1981). The complete amino aid sequence of α-neo-endorphin. *Biochem. Biophys. Res. Commun.* 99, 871–878.

Kilpatrick D. L., Lewis, R. V., Stein, S., and Udenfriend, S. (1980). Release of enkephalins and enkephalin-containing polypeptides from perfused beef adrenal glands. *Proc. Natl. Acad. Sci. USA* 77, 7473–7475.

Kilpatrick, D. L., Taniguchi, T., Jones, B. N., Stern, A. S., Shively, J. E., Hullihan, J., Kimura, S., Stein, S., and Udenfriend, S. (1981a). A highly potent 3200-dalton adrenal opioid peptide that contains both a [Met]- and [Leu]enkephalin sequence. *Proc. Natl. Acad. Sci. USA* 78, 3265–3268.

Kilpatrick, D. L., Jones, B. N., Kojima, K., and Udenfriend, S. (1981b). Identification of the octapeptide [Met] enkephalin-Arg⁶–Gly⁷–Leu⁸ in extracts of bovine adrenal medulla. *Biochem. Biophys. Res. Commun. 103*, 698–705.

Kilpatrick, D. L., Jones, B. N., Lewis, R. V., Stern, A. S., Kojima, K., Shively, J. E., and Udenfriend, S. (1982a). An 18 200-dalton adrenal protein that contains four [Met] enkephalin sequences. *Proc. Natl. Acad. Sci. USA 79*, 3057–3061.

Kilpatrick, D. L., Wahlström, A., Lahm, H. W., Blacher, R., Ezra, E., Fleminger, G., and Udenfriend, S. (1982b). Characterization of rimorphin, a new [Leu] enkephalin-containing peptide from bovine posterior pituitary glands. *Life Sci. 31*, 1849–1852.

Kilpatrick, D. L., Wahlstrom, A., Lahm, H. W., Blacher, R., and Udenfriend, S. (1982c). Rimorphin, a unique, naturally occurring [Leu] enkephalin-containing peptide found in association with dynorphin and α-neo-endorphin. *Proc. Natl. Acad. Sci. USA 79*, 6480–6483.

Kilpatrick, D. L., Howells, R. D., Lahm, H.-W., and Udenfriend, S. (1983a). Evidence for a proenkephalin-like precursor in amphibian brain. *Proc. Natl. Acad. Sci. USA 80*, 5772–5775.

Kilpatrick, D. L., Eisen, M., Ezra, E., and Udenfriend, S. (1983b). Processing of prodynorphin at single and paired basic residues in porcine neurointermediate lobe. *Life Sci. 33*, Suppl. 1, 93–96.

Kimura, S., Lewis, R. V., Gerber, L. D., Brink, L., Rubinstein, M., Stein, S., and Udenfriend, S. (1979). Purification to homogeneity of camel pituitary pro-opiocortin, the common precursor of opioid peptides and corticotropin. *Proc. Natl. Acad. Sci, USA 76*, 1756–1759.

Kimura, S., Lewis, R. V., Stern, A. S., Rossier, J., Stein, S., and Udenfriend, S. (1980). Probable precursors of [Leu] enkephalin and [Met] enkephalin in adrenal medulla: Peptides of 3–5 kilodaltons. *Proc. Natl. Acad. Sci. USA 77*, 1681–1685.

Kitamura, K., Minamino, N., Hayashi, Y., Kangawa, K., and Matsuo, H. (1982). Regional distribution of β-neo-endorphin in rat brain and pituitary. *Biochem. Biophys. Res. Commun. 109*, 966–974.

Kojima, K., Kilpatrick D. L., Stern, A. S., Jones, B. N., and Udenfriend, S. (1982). Proenkephalin: a general pathway for enkepahlin biosynthesis in animal tissues. *Arch. Biochem. Biophys. 215*, 638–643.

Lewis, R. V. and Stern, A. S. (1983). Biosynthesis of the enkephalins and enkephalin-containing polypeptides. *Ann. Rev. Pharmacol. Toxicol. 23*, 353–372.

Lewis, R. V., Stein, S., Gerber, L. D., Rubinstein, M., and Udenfriend, S. (1978). High molecular weight opioid-containing proteins in striatum. *Proc. Natl. Acad. Sci. USA 75*, 4021–4023.

Lewis, R. V., Stern, A. S. Rossier, J., Stein, S., and Udenfriend, S. (1979). Putative enkephalin precursors in bovine adrenal medulla. *Biochem. Biophys. Res. Commun. 89*, 822–829.

Lewis, R. V., Fallon, A., Stein, S., Gibson, K. D., and Udenfriend, S. (1980). Supports for reverse-phase high-performance liquid chromatography of large proteins. *Anal. Biochem. 104*, 153–159.

Lewis, R. V., Stern, A. S., Kimura, S., Rossier, J., Stein, S., and Udenfriend, S. (1980b). An about 50 000-dalton protein in adrenal medulla: a common precursor of [Met]- and [Leu]-enkephalin. *Science 208*, 1459–1461.

Lewis, R. V., Stern, A. S., Kimura, S., Stein, S., and Udenfriend, S. (1980c). Enkephalin biosynthetic pathway: proteins of 8000 and 14 000 daltons in bovine adrenal medulla. *Proc. Natl. Acad. Sci. USA 77*, 5018–5020.

Li, C. H., Barnafi, L., Chrétien, M. and Chung, D. (1965). Isolation and amino-acid sequence of β-LPH from sheep pituitary glands. *Nature 208*, 1093–1094.

Lindberg, I., Yang, H.-Y. T., and Costa, E. (1982). An enkephalin-generating enzyme in bovine adrenal medulla. *Biochem. Biophys. Res. Commun. 106*, 186–193.

Ling, N., Burgus, R., and Guillemin, R. (1976). Isolation, primary structure, and synthesis of α-endorphin and γ-endorphin, two peptides of hypothalamic-hypophysial origin with morphinomimetic activity. *Proc. Natl. Acad. Sci. USA 73*, 3942–3946.

Livett, B. G., Dean, D. M., Whelan, L. G., Udenfriend, S., and Rossier, J. (1981). Co-release of enkephalin and catecholamines from cultured adrenal chromaffin cells, *Nature 289*, 317–319.

Minamino, N., Kangawa, K., Chino, N., Sakakibara S., and Matsuo, H. (1981). β-Neo-endorphin, a new hypothalamic 'big' leu-enkephalin of porcine origin: its purification and the complete amino acid sequence. *Biochem. Biophys. Res. Commun. 99*, 864–870.

Mizuno, K., Minamino, N., Kangawa, K. and Matsuo, H. (1980a). A new endogenous opioid peptide from bovine adrenal medulla: isolation and amino acid sequence of a dodecapeptide (BAM-12P). *Biochem. Biophys. Res. Commun. 95*, 1482–1488.

Mizuno, K., Minamino, N., Kangawa, K., and Matsuo, H., (1980b). A new family of endogenous 'big' met-enkephalins from bovine adrenal medulla: purification and structure of docosa- (BAM-22P) and eicosapeptide (BAM-20P) with very potent opiate activity. *Biochem. Biophys. Res. Commun. 97*, 1283–1290.

Mizuno, K., Miyata, A., Kangawa, K., and Matsuo, H. (1982). A unique proenkephalin-convert- ing enzyme purified from bovine adrenal chromaffin granules. *Biochem. Biophys. Res. Commun. 108*, 1235–1242.

Nakanishi, S., Inoue, A., Kita, T., Nakamura, M., Chang, A. C. Y., Cohen, S. N., and Numa, S. (1979). Nucleotide sequence of cloned cDNA for bovine corticotropin-β-lipotropin precursor. *Nature 278*, 423–427.

Noda, M., Furutani, Y., Takahashi, H., Toyosato, M., Hirose, T., Inayama, S., Nakanishi, S., and Numa, S. (1982). Cloning and sequence analysis of cDNA for bovine adrenal preproenkephalin. *Nature 295*, 202–206.

Orlowski, H. and Wilk, S. (1981). A multicatalytical protease complex from pituitary that forms enkephalin and enkephalin containing peptides. *Biochem. Biophys. Res. Commun. 101*, 814–822.

Possenti, R., De Marco, V., Cherubini, O., and Roda, L. G. (1983). Enkephalin-binding systems in human plasma. *Neurochem. Res. 8*, 423–432.

Przewtocki, R., Gramsch, C., Pasi, A., and Herz, A. (1983). Characterization and localization of immunoreactive dynorphin, α-neo-endorphin, met-enkephalin and substance P in human spinal cord. *Brain Res. 280*, 95–103.

Rossier, J., Trifaró, J. M., Lewis, R. V., Lee, R. W. H., Stern, A., Kimura, S., Stein, S., and Udenfriend, S. (1980). Studies with [^{35}S]methionine indicate that the 22 000-dalton [Met]enkephalin-containing protein in chromaffin cells is a precursor of [Met]enkephalin. *Proc. Natl. Acad. Sci. USA 77*, 6889–6891.

Rossier, J., Dean, D. M., Livett, B. G., and Udenfriend, S. (1981). Enkephalin congeners and precursors are synthesized and released by primary cultures of adrenal chromaffin cells. *Life Sci. 28*, 781–789.

Rubinstein, M., Stein, S., Gerber, L. D., and Udenfriend, S. (1977a). Isolation and characteri- zation of the opioid peptides from rat pituitary: β-Lipotropin. *Proc. Natl. Acad. Sci. USA 74*, 3052–3055.

Rubinstein, M., Stein, S., and Udenfriend, S. (1977b). Isolation and characterization of the opioid peptides from rat pituitary: β-Endorphin. *Proc. Natl. Acad. Sci. USA 74*, 4969– 4972.

Rubinstein, M., Stein, S., and Udenfriend, S. (1978). Characterization of pro-opiocortin, a precursor to opioid peptides and corticotropin. *Proc. Natl. Acad. Sci. USA 75*, 669–671.

Sabol, S. L., Liang, C.-M., Dandekar, S., and Kranzler, L. S. (1983). *In vitro* biosynthesis and processing of immunologically identified methionine-enkephalin precursor protein. *J. Biol. Chem. 258*, 2697–2704.

Schwartz, J.-C., Malfroy, B., and De La Baume, S. (1981). Biological inactivation of enkephalins and the role of enkephalin-dipeptidyl-carboxypeptidase ('enkephalinase') as Neuropeptidase. *Life Sci. 29*, 1715–1740.

Seizinger, B. R., Grimm, C., Höllt, V., and Herz, A. (1984). Evidence for a selective processing of proenkephalin B into different opioid peptide forms in particular regions of rat brain and pituitary. *J. Neurochem. 42*, 447–457.

Stein, S., Böhlen, P., Stone, J., Dairman, W., and Udenfriend, S. (1973). Amino acid analysis with fluorescamine at the picomole level. *Arch. Biochem. Biophys. 155*, 203–212.

Stern, A. S., Lewis, R. V., Kimura, S., Rossier, J., Gerber, L. D., Brink, L., Stein, S., and Udenfriend, S. (1979). Isolation of the opioid heptapeptide Met-enkephalin[Arg^6, Phe^7]

from bovine adrenal medullary granules and striatum. *Proc. Natl. Acad. Sci. USA 76*, 6680–6683.

Stern, A. S., Lewis, R. V., Kimura, S., Rossier, J., Stein, S., and Udenfriend, S. (1980). Opioid hexapeptides and heptapeptides in adrenal medulla and brain: Possible implications on the biosynthesis of enkephalins. *Arch. Biochem. Biophys. 205*, 606–613.

Stern, A. S., Jones, B. N., Shively, J. E., Stein, S., and Udenfriend, S. (1981). Two adrenal opioid polypeptides: Proposed intermediates in the processing of proenkephalin. *Proc. Natl. Acad. Sci. USA 78*, 1962–1966.

Suda, T., Tozawa, F., Tachibana, S., Demura, H., and Shizume, K. (1982). Multiple forms of immunoreactive dynorphin in rat pituitary and brain. *Life Sci. 31*, 51–57.

Tachibana, S., Araki, K., Ohya, S., and Yoshida, S. (1982). Isolation and structure of dynorphin, an opioid peptide, from porcine duodenum. *Nature 295*, 339–340.

Tanaka, S., Oshima, T., Ohsue, K., Ono, T., Oikawa, S., Takano, I., Noguchi, T., Kangawa, K., Minamino, N., and Matsuo, H. (1982). Expression in *Escherichia coli* of chemically synthesized gene for a novel opiate α-neo-endorphin. *Nucleic Acids Res. 10*, 1741–1754.

Tang, J., Yang, H.-Y. T., and Costa, E. (1982). Distribution of Met[5]-enkephalin-Arg[6]–Phe[7] (MEAP) in various tissues of rats and guinea pigs. *Life Sci. 31*, 2303–2306.

Tang, J., Chou, J., Zhang, A. Z., Yang, H.-Y. T., and Costa, E. (1983). Met[5]-enkephalin-Arg[6]–Phe[7] and its receptor in lung. *Life Sci. 32*, 2371–2377.

Troy, C. M. and Musacchio, J. M. (1982). Processing of enkephalin precursors by chromaffin granule enzymes. *Life Sci. 31*, 1717–1720.

Udenfriend, S. and Kilpatrick, D. L. (1983). Biochemistry of the enkephalins and enkephalin-containing peptides. *Arch. Biochem. Biophys. 221*, 309–323.

Udenfriend, S., Kilpatrick, D. L., and Kojima, K. (1983). High performance liquid chromatography-fluorometric determination of primary amines, peptides and proteins: Picomole level detection with fluorescamine In: *Methods in Biogenic Amine Research*. S. Parvez, T. Nagatsu, I. Nagatsu, and H. Parvez, (eds). Elsevier, Amsterdam (pp. 1–12).

Weber, E., Evans, C. J., Chang, J.-K., and Barchas, J. D. (1982). Brain distributions of α-neo-endorphin and β-neo-endorphin: Evidence for regional processing differences. *Biochem. Biophys. Res. Commun. 108*, 81–88.

Weber, E., Esch, F. S., Böhlen, P., Paterson, S., Corbett, A. D., McKnight, A. T., Kosterlitz, H. W., Barchas, J. D., and Evans, C. J. (1983). Metorphamide: Isoltion, structure, and biologic activity of an amidated opioid octapeptide from bovine brain. *Proc. Natl. Acad. Sci. USA 80*, 7362–7366.

Progress in HPLC, Vol. 1, pp. 179–217
Parvez *et al.* (Eds)
© 1985 VNU Science Press

Technical aspects of biochemical high-performance column liquid chromatography

JÖRGEN SJÖDAHL,[1] KEITH J. DILLEY,[1] ROLF ERIKSSON,[1] JOEL
PELLERIN,[2] S. H. PARVEZ,[3] and LENNART ARLINGER[1]

[1] LKB-Produkter AB, Box 305, S-161 26 Bromma, Sweden, [2] LKB Instruments
S. A., Boite Postale 29, F-91404 Orsay Cedex, France, and [3] Unité de
Pharmacologie, Université de Paris XI, Centre D'Orsay – Bat. 440, 91405 Orsay
Cedex, France

INTRODUCTION

Historical overview
Readers will be aware of the tremendous growth in high-performance liquid
chromatography (HPLC) applications in biochemical studies. The reason for this
rapid and recent development is primarily related to the availability of new high-
performance packing materials. These materials are optimized for the separation of
biopolymers such as proteins, nucleic acids, polysaccharides, and their fragments
by 'gel filtration', 'ion-exchange', 'hydrophobic interaction', or 'reversed-phase'
chromatography. Regnier and Gooding predicted in 1980 that by the end of
the decade 70–80% of all protein purification would be achieved by high or
medium pressure liquid chromatography – a prediction based upon the assumed
rapid acceptance of new packing materials.

Before discussing the specific technical requirements for high-performance
separation of biomolecules, it is relevant to briefly summarize the development
of HPLC as we know it today. For a detailed historical review the reader is referred
to the excellent overview of the evolution of the technique compiled by Ettre
(1971). Laying emphasis on technical aspects, some of the turning points identified
are not very often referred to.

Fractionation of crude oil on pulverized Fuller's earth performed by David
Talbot Day in 1897 is the earliest known example of liquid chromatography.
The first biochemical application was described in 1903 when the Russian scientist
Michael Tswett separated vegetable pigments on calcium carbonate. Tswett realized
the general applicability of the concept, including the importance of adapting the
sample to the only powerful detection device available at the time – the human
eye! Partition chromatography was developed approximately 40 years later (Martin
and Synge, 1941), and this in turn led to the development of gas chromatography
instrumentation.

In 1948, Moore and Stein described the basis for a semiautomated system for the quantitation of amino acids after separation by ion-exchange chromatography: the amino acid analyzer. In our opinion this represents the birth of the first HPLC system, which has evolved into the instrumentation we have today. Moore's and Stein's later HPLC system included such features as programmable solvent delivery, semiautomatic injection, temperature programming, post-column derivatization, and dual-wavelength on-line detection.

For the chromatographic separation of biological macromolecules, the break-through came in 1959, when Porath and Flodin developed the cross-linked polydextran gels (later commercially available as Sephadex). Together with the development of derivatized cellulose for ion-exchange chromatography (Peterson and Sober, 1956), biochemists now had available some very effective tools for purification of macromolecules. Shortly after the introduction of these new matrices, a range of chromatographic instrumentation optimized for soft gel (large particle) liquid chromatography became available.

The development of column support materials and instrumentation was paral-lelled by attempts to describe the chromatographic separation process from a theoretical standpoint. This culminated in a classical presentation of chromato-graphic theory (Giddings, 1965).

The need for faster chromatography, combined with higher resolution was the driving force behind the development of new packing materials and instrumentation. As a result, the performance standards for the separation of low molecular weight compounds were significantly improved in the late 1960s by the independent appearance of small diameter, rigid, column packing materials as well as improved dedicated equipment for ion exchange (the amino acid analyzer) and reversed/normal-phase ('conventional' HPLC) chromatography.

Parris (1976) made a review on instrumental aspects of 'conventional' (nonbio-chemical) HPLC which is recommended to those inexperienced in the technique. Figure 1 shows the basic system components and configuration in a system for current high-performance column liquid chromatography. Note the large number of instrument options that are now available, particularly for detectors and output devices.

For the separation of biological macromolecules, however, soft gel materials exhibiting medium performance were the only choice until the introduction in 1979 (by Toyo Soda Co., Japan) of high-performance columns prepacked with a biochemically compatible silica matrix for high-performance gel filtration. This made rapid, high resolution separation of biological macromolecules possible for the first time — if one defines 'rapid' according to today's standards. In addition, the work of Chang et al. (1976) provided a basis for the development of high-performance ion exchangers for the separation of biopolymers. Columns for this rapidly growing technique are now available from several manufacturers.

The appearance of these high-performance columns for biomacromolecular separations in the late 1970s, created a situation similar to that which had occurred previously in the early 1960s for soft gels; the separation media were available but the optimized instrumentation was missing. As the stationary phase can be considered to be the vital component in all types of chromatography, it follows

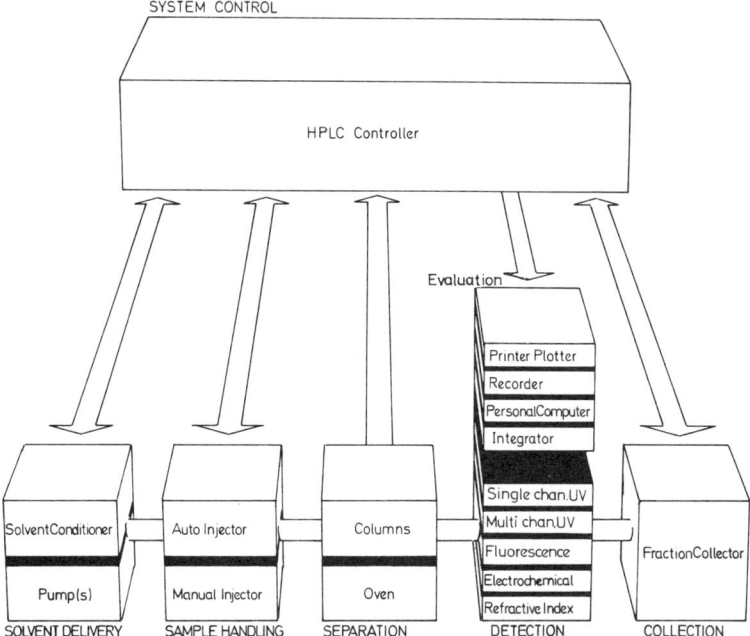

SYSTEM CONTROL

HPLC Controller

Evaluation

Printer Plotter	
Recorder	
PersonalComputer	
Integrator	

SolventConditioner	Auto Injector	Columns	Single chan.UV	FractionCollector
			Multi chan.UV	
			Fluorescence	
			Electrochemical	
Pump(s)	Manual Injector	Oven	Refractive Index	
SOLVENT DELIVERY	SAMPLE HANDLING	SEPARATION	DETECTION	COLLECTION

Figure 1. Column liquid chromatography, some system configurations.

that the instrumentation constitutes only a tool to utilize the separation occurring in the column. Even so, the design of the instrumentation influences the performance of the system to a very significiant extent. Thus, it is important to realize that equipment for what is normally termed 'HPLC' in the period up to 1980 was optimized exclusively for the separation of small molecules or organic polymers in predominantly nonaqueous solvent systems.

In the remainder of this chapter we will describe instrumentation optimized for the separation of biomolecules, especially biomacromolecules. Presently, three main categories of instrumentation are being used for separations with the new high-performance biomacromolecule columns, as given in Table 1. This explains why a new breed of biochemical HPLC instrumentation has evolved, as a fusion of ideas from conventional HPLC instrumentation and the instrumentation used for purification of biomacromolecules on soft gels.

Biochemical HPLC: some definitions
Before discussing technical aspects, we feel it is important to comment on the nomenclature used in both biochemical and analytical chemical HPLC. Many terms are currently in use, both new and old, and the frequent non-systematic use of these terms often creates confusion, particularly for newcomers to the technique.

Chromatography is a technique for separating compounds by exploiting differences in their interaction with a stationary phase when transported by a second, mobile phase and simultaneously allowed to interact. It is possible to obtain in

Table 1. HPLC instrumentation categories, laboratory scale

System	Applicability
Conventional HPLC instrumentation	Excellent for non-biochemical HPLC, problems with optimization and reliability in the biochemical mode
Biochemical MPLC[a] instrumentation	Excellent for most (low to medium pressure) biochemical HPLC separations but cannot be used for many conventional HPLC applications mainly due to pressure limitations
Biochemical HPLC instrumentation	Excellent for the majority of all current HPLC applications

a Medium Pressure Liquid Chromatography

theory nine (if we exclude intermediates such as supercritical fluids) submodes of chromatography ranging from gas-gas through gas-liquid, ... , solid-liquid to solid-solid-chromatography. However, it has to be admitted that the above extremes (gas-gas and solid-solid) have not to our knowledge been realized in any practical system. In liquid chromatography, liquid-liquid or solid-liquid chromatography is used. Thus for the stationary phase, both solids and liquids immobilized to solids are used. If a liquid is used as a stationary phase and if this liquid is not immobilized to a solid, but still physically dispersed from the mobile liquid due to non-miscibility, the technique is called 'two-phase partitioning'.

If a porous particle is used in combination with the liquid mobile phase, and the pore size is utilized to control the interaction (sterically hindered diffusion) of the sample compounds with the stationary phase (the pores), the common names are 'gel filtration', 'gel permeation', 'size exclusion', or 'steric exclusion chromatography' — 'gel filtration' traditionally implying soft particles, 'gel permeation' implying rigid particles irrespective whether these are gels or not.

In column liquid chromatography, the mobile phase is, thus, a liquid and the stationary phase is packed into a column and retained by the inlet/outlet frits; alternatively, the stationary phase can be chemically bonded to the wall of the column (Nota et al., 1970).

If we concentrate on packed columns, depending on the size of the particles and/or the column, the linear flow rate, or the application, 'very high-speed LC' and 'low dispersion LC' are examples of unnecessary attempts to create new names for minor variants of previously established techniques. The same trend can also be seen with the abbreviation FPLC (fast protein liquid chromatography), which has been used to describe the separation of bio-molecules by high-performance 'ion-exchange' chromatography and recently by high-performance 'reversed-phase' chromatography. These comments perhaps serve to indicate the urgent need for chemists and biochemists to start a joint discussion to unify the nomenclature — the methodologies used by these groups of scientists now show considerable overlap.

In the high-performance mode, i.e. when using rigid particles with small particle size in combination with the appropriate instrumentation, the user has the possibility to select the level of performance for any particular application. Overall performance will be defined by the combination of the requirements for (1) resolution, (2) separation time, and (3) capital and operating costs. We consider technical aspects of biochemical HPLC to be the implications and optimization requirements generated by the need for acceptable performance (as discussed) from the new generation of HPLC columns for biomolecules.

This chapter will review aspects of instrument optimization for the separation of biomolecules in analytical, micropreparative as well as laboratory preparative scale using packed columns for 'gel' filtration', 'ion-exchange', 'hydrophobic inter-action' chromatography including 'reversed-phase' chromatography, and 'affinity' chromatography. These areas of column liquid chromatography are represented by the 'east' portion of the 'column liquid chromatography space', shown in Fig. 2. This figure shows how different applications fit into the overall matrix for LC.

DESIGN ASPECTS OF BIOCHEMICALLY OPTIMIZED HPLC-INSTRUMENTATION

General requirements of the biochemist

When considering HPLC for the biochemist, the technique should not be viewed in isolation. Indeed, before proceeding to discuss HPLC we will briefly mention

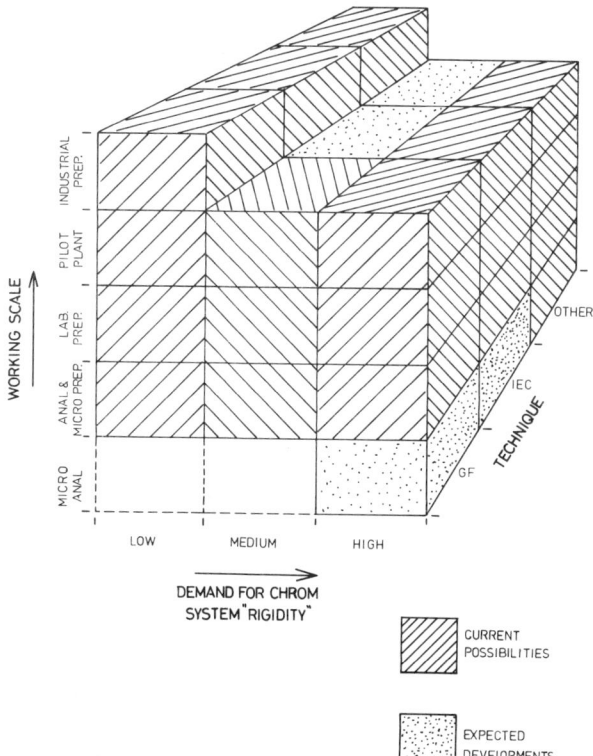

Figure 2. An overall 3D matrix for column liquid chromatography. The omitted 'cubes' are considered to be of minor importance, either due to reduced application interest (lower parts) or in most cases unacceptable costs (upper parts).

184   *J. Sjödahl et al.*

another separation technique available to the biochemist: electrophoresis. This is important because it can provide valuable information before and after high-performance liquid chromatographic separations. Too often, in our opinion, the criteria 'pure by HPLC' is used for a single peak emerging from a system. The statement 'pure by HPLC and electrophoresis' has far greater validity.

There are already several excellent reviews of the instrumentation used in electrophoretic techniques, for example, Deyl (1979). In the context of the HPLC separation of biomacromolecules, gel-electrophoresis can perform the following invaluable functions:

1. *Determination of sample complexity*: two-dimensional (2D) separation is one of the highest resolving techniques available to the biochemist. The complex information obtained (see Fig. 3 for example) provides proof of the frequent need to use high-performance separation techniques whether electrophoretic or chromatographic.

Isoelectric focusing (IEF) and SDS-polyacrylamide gel electrophoresis (SDS–PAGE) are the techniques commonly combined into the two-dimensional mode. They can, of course, be used independently to determine sample heterogeneity with respect to charge and size, respectively, before the chromatographic separation.

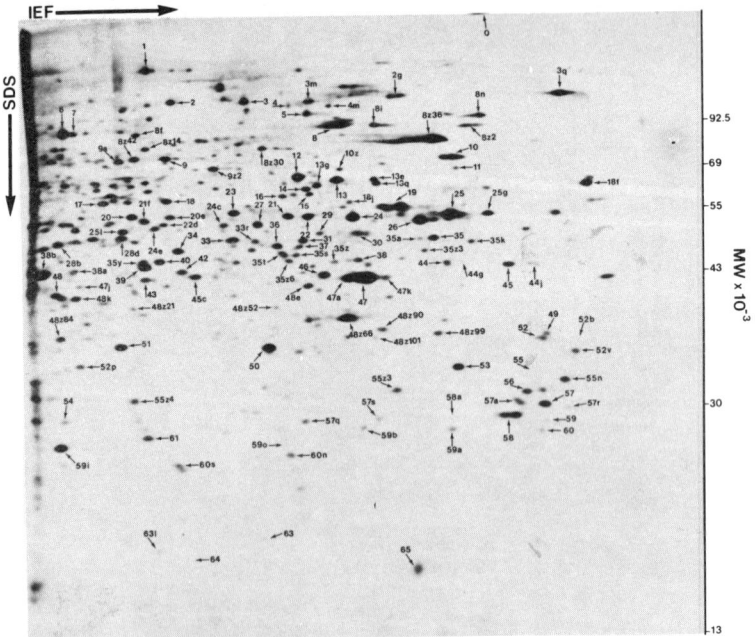

Figure 3. Two-dimensional gel electrophoresis map of HeLa acidic radioactively labelled proteins. [Courtesy of Rodrigo Bravo (EMBL, Heidelberg) and Julio Celis (Department of Chemistry, Aarhus University) from *Clin. Chem.* 28, 766–781 (1982).]

2. *Analysis of titration curves* is an interesting electrophoretic application which reveals the net charge/pH relationship for components in a protein mixture (Rosengren *et al.*, 1976). Knowledge of the titration curve for a protein mixture can greatly simplify establishing the conditions for purification of the sample by high-performance 'ion-exchange' chromatography. The information obtained will determine whether to use a cation or anion exchanger and the pH of the mobile phase for optimal charge differences between sample components. This relatively simple technique can, in our opinion, remove much of the 'trial and error' that has previously characterized attempts to determine optimal conditions for separations based on 'ion-exchange' chromatography.

3. *Determination of peak purity*: collection of an eluted compound and subsequent analysis by an electrophoretic method such as immunoelectrophoresis will either give confirmation of purification, or reveal the need for further chromatography steps.

Considering chromatography, it is worth remembering that biochemists have for decades achieved a reasonable quality of separation by using the available soft gel media. However, the availability of rigid and biochemically compatible packing materials permits much smaller particles to be used. Chromatography, being a process of multiple diffusion at equilibria, will benefit by all attempts to improve mass transfer between the mobile and the stationary phases. The use of small particles is one very important way of improving the mass transfer by reducing the physical distances a solute must cover when diffusing.

By the use of small-particle packing materials, the new biochemical HPLC columns give a great improvement in both resolution and separation time. Today, 'gel filtration', 'ion-exchange', and to some extent also 'reversed-phase' chromatography, are extensively used in macromolecular HPLC separations. 'Reversed-phase' chromatography is one of the most powerful methods for HPLC-separations of medium-sized to small biomolecules presently available. In the future such molecules will be frequently separated also by complementary techniques such as 'ion-exchange' and 'hydrophobic interaction' chromatography.

If soft gels give inferior separation and run times are longer, then why should the biochemist still use them? There are a number of reasons such as broader range, easier scale-up, better documented methods for 'affinity' chromatography, and last but not least, cost. However, most if not all, of these advantages are expected to disappear in the near future. In addition, the high-performance methods involve less manual labour and are inherently more reliable than the corresponding soft gel techniques. Large scale chromatography on the industrial scale appears to be the only major area where soft gels will continue to be cost effective.

As already stated in the Introduction, equipment optimized for analytical applications of HPLC has been available for over a decade. There are, however, some key differences in the way a biochemist and an analytical chemist need to use their HPLC systems to optimal effect. Very generally speaking, the analyst usually only determines the areas of the separated peaks, whereas the biochemist often collects the purified components. In addition to their different interests in purification or analysis, the two disciplines also show very considerable differences in sample type. The analyst's samples usually consist of relatively simple, non-polar, low molecular weight components; typically a result of an efficient sample pretreatment. In contrast, the biochemist's samples usually consist of complex, relatively polar, and often high-molecular weight components. Some of these differences are shown in Fig. 4.

In addition, the requirement of the biochemist to separate very complex samples without the usual 'analytical' approach of creating a 'separation window' (conditions focusing on the separation of a few interesting compounds, eluting non-interesting compounds non-separated) will necessitate the more frequent use of complex gradient elution techniques. Furthermore, for sample injection, there is a main difference between 'conventional' HPLC and 'biochemical' HPLC: the need in the latter application area to inject large volumes of sample in the cases where the pre-concentration of biomolecules is difficult or impossible.

To summarize, the identifiable differences between 'biochemical' HPLC and 'conventional analytical' HPLC are:

- new requirements for solvent delivery systems;
- an increased use of complex gradients for eluting the sample components;
- an increased need to repeatedly inject large volumes onto columns of low capacity (high-capacity columns would perhaps be cost-prohibitive); and
- the frequent use of eluate collection and the need for a high performance/resolution capability in eluate collection.

These points, which are specifically 'biochemically induced' will be discussed in detail. Further points which arise by new trends within the technical development of HPLC as such will be treated in the following section with special reference to the impact such trends are expected to have on 'biochemical' HPLC.

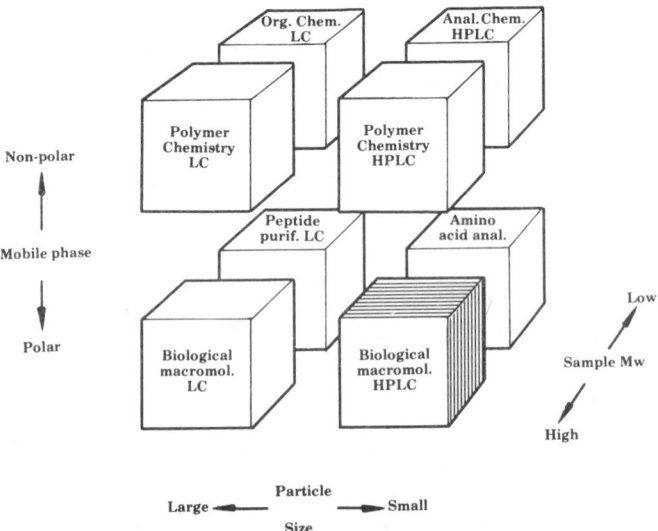

Figure 4. Biochemical vs analytical chemical liquid chromatography. The figure shows the relative positioning of different types of column liquid chromatography: biochemical HPLC vs analytical chemistry HPLC is one example.

Solvent delivery

Biomacromolecules. As can be seen in Fig. 4 the sample size/molecular weight is one of the parameters differentiating 'analytical' HPLC from 'biochemical' HPLC. The main effect of this from an instrumental point of view is related to the solvent delivery system: the flow rate range to be covered by the pump (Sjödahl *et al.*, 1982a, b), the ability to function reliably when operating at very low back pressures, and the reliability problems commonly observed when running aqueous polar solvents and high molarity buffers through a conventional HPLC pump. Problems of the last category are illustrated in Fig. 5. The requirement for an extended flow rate range can be established by consideration of the influence of sample molecular weight on the choice of flow rate. To reach this conclusion it is necessary to review the previously emphasized (Sjödahl, 1980) relationship between separation parameter optimization and sample molecular size from a theoretical point of view. For a detailed theoretical treatment of general chromatography, see Giddings (1965).

In column chromatography the separation of two compounds, A and B, is due to their different distribution between the mobile phase and the stationary phase. The resulting separation, the resolution, depends on two parameters: the *selectivity* of the system, reflected by the volumetric distance between the maxima of the two eluted peaks, and the *band-broadening* effects of the entire separation system, indicated by the widths of the peaks. The resolution is defined as $R = 2V/(W_A + W_B)$, where V is the volumetric distance between the peaks, and W_A and W_B refer to the baseline volumetric widths of the two peaks of compounds A and B, respectively. In

Figure 5. Corrosion of a conventional HPLC pump head. The photograph was taken after six months use of a conventional HPLC pump in biochemical applications.

order to increase the separation efficiency of a system it is thus possible to alter parameters affecting either the selectivity or the band-broadening effects, or both.

Selectivity. In spite of the absence of high-efficiency packing materials, excellent separations could be achieved decades ago. This was possible because the poor band-broadening characteristics of traditional packing materials were compensated for by a reasonable selectivity of the systems used. By utilizing various types of chromatographic interaction the distance (volume) between peaks was often adequate.

In the field of liquid chromatography, biomacromolecules can be separated according to size, net surface charge, hydrophobicity, and three-dimensional substructures, by using 'gel filtration', 'ion-exchange' chromatography, non-ionic adsorption chromatography, and biospecific-interaction ('affinity') chromatography. In the latter three techniques many possibilities exist to alter the selectivity, either by introducing various interacting groups onto the surface of the stationary phase, or simply by changing the composition of the eluent. In 'gel filtration', in contrast, the selectivity is dependent solely on the properties of the stationary phase.

Selectivity in true 'gel filtration' is, by definition, governed by the pore size and the porosity of the stationary phase: the difference in eluent volume between the first and the last eluted peaks is always less than or equal to the pore volume of the beads. It should also be noticed that the maximal distance between the first peak and the last one is always less than the column volume. In other words, the only way to achieve better separation in 'gel filtration' is to reduce the band-broadening effects; control of the selectivity of existing gels has reached its potential limit. In other types of liquid chromatographic techniques, it is possible both to reduce the band-broadening effects and to further improve the selectivity.

Band-broadening effects. First of all, we shall assume that a good instrumental set up is used, where the effects of band broadening in the injection system, the detector flow cell, and the connecting capillaries have all been minimized. Band broadening in the column can be attributed to three principal diffusion effects (Giddings, 1965): (1) eddy diffusion, reflecting how well the column has been packed. The presence of voids and/or irregularly shaped particles greatly increases the magnitude of this effect; (2) axial diffusion, depending on both the diffusion coefficient of the sample and on the residence time of the sample in the mobile phase; and (3) dispersion due to incomplete mass transfer, reflecting non-equilibrium conditions for interaction between the sample and the stationary phase. The summation of these three effects will be a more or less pronounced broadening of the peaks. The efficiency of a column with respect to band broadening is often expressed as the number of theoretical plates, N

$$N \sim 5.54(V_e/W_{0.5})^2$$

where V_e is the elution volume of a peak, and $W_{0.5}$ is the width of the peak at half its height.

In order to compare columns of different lengths, the height equivalent to a theoretical plate, H, is calculated as the quotient of the column length, l, and the number of theoretical plates: $H = l/N$. A small value for H is consistent with a high-separation efficiency. It has been shown theoretically (Giddings, 1965) that the phenomena described above, eddy diffusion, axial diffusion, and dispersion due to incomplete mass transfer, add up to a resulting band broadening according to the equation

$$H = A + B/u + Cu$$

where u equals the linear flow velocity of the eluent through the column; A reflects the eddy diffusion, and is proportional to the particle size, d_p, of the stationary phase; B is proportional to the diffusion coefficient, D_m, of the sample molecules in the mobile phase: the term B/u reflects the axial diffusion; and C is related to both the particle size and the diffusion coefficient (and, also to the diffusion in the stationary phase).

The relationship can be approximated to

$$H \sim k_1 d_p + k_2 D_m/u + k_3 d_p{}^2 u/D_m$$

where k_1, k_2, and k_3 are constants. Analysing this expression, and remembering that a small value of H means a high separation efficiency, it is seen that:

(a) a small particle size, d_p, is consistent with a generally improved efficiency, and a reduction of the negative effects of excessive flow rates on the separation efficiency;

(b) a low diffusion coefficient, D_m, reduces the effect of axial diffusion (which, on the other hand, is not so important at practical flow rates), as well as increases the negative effects of high flow rates; and

(c) an optimal flow rate exists, the value of which is related to the particle size d_p as well as to the diffusion coefficient D_m of the sample (related to sample molecular weight).

Due to these theoretical considerations and the unfortunate circumstances that 'gel filtration' matrices are (1) commonly used for separation of high-molecular weight compounds; and (2) not easily improved by increasing mass-transfer through a further decrease of particle size, a further major improvement of the separation possibilities offered by this mode of HPLC is difficult to envisage. Flow rate will continue to be of great importance in the separation of high-molecular weight samples of 'gel filtration' as shown in Fig. 6. For other separation methods, such as, 'ion-exchange' chromatography, the same effects are also observed, but (1) the possibility of improving mass-transfer by utilizing, more or less, only the outer surface of the beads; (2) the common use of this method for small or medium-sized (i.e. non-large) molecules; and (3) the higher selectivity, not always necessitating optimization of band-broadening, all lessen the importance of flow-optimization in 'non-gel filtration' separations, especially in applications where the molecules of interest are small to medium-sized.

The practical result of the theoretical treatment discussed above is that it is very important to know the flow rate vs efficiency function for the columns used in a particular application: a noticeable effect will be observed, the magnitude of which will increase with the molecular weight of the sample especially if 'gel-filtration' methods are used.

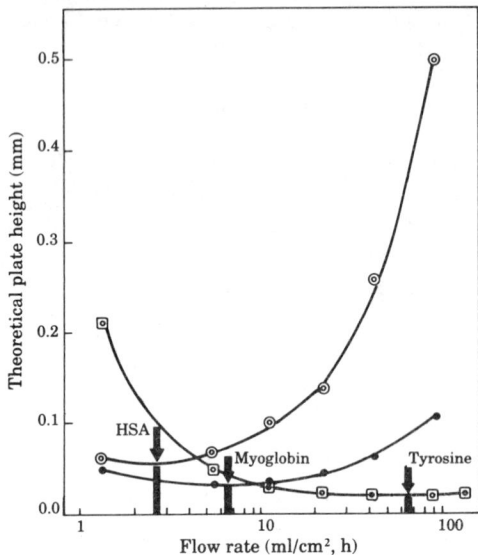

Figure 6. The importance of flow rate in high performance 'gel filtration'. The relationship between separation efficiency (expressed as height equivalent to a theoretical plate) and linear flow rate for substances of different molecular weight. Albumin (mol. wt 68 000); myoglobin (mol. wt 17 000); tyrosine (mol. wt 180). Arrows indicate optimum flow rate for the respective substance. Column: LKB 2135 Ultropac TSK SW3000, 7.5 × 600 mm. Sample: 200 μl of a 1% (w/v) solution of the respective substance in 1 mM Na_2HPO_4, 0.1 M KCl, 0.05% NaN_3 (w/v), pH 7.4.

The importance of flow optimization in 'gel-filtration' HPLC is further shown in Fig. 7 which shows human serum proteins separated at different flow rates. It can be debated which of the results represents the highest performance: we would say all three, depending on the specific need in a particular case. The particular conditions selected have to be a compromise between the time available and the degree of resolution required.

It is interesting to note that, due to the low working pressure generated at very low flow rates the 0.01–0.1 ml/min results shown in Fig. 7 can be achieved by using peristaltic pumps. The medium to high flow rate results are accessible with 'conventional' HPLC pumps, however, lower flow rate results are not.

The optimal situation is, of course, to be able to select from a menu spanning all three conditions by utilizing modern general-purpose solvent delivery systems.

Biochemical conditions. Separation of biological molecules are usually performed at relatively high ionic strengths. The effect of such solutions (which often contain chloride ions) on a metal part of a typical conventional HPLC pump is shown in Fig. 5. As can be seen, severe corrosion of essential components can occur. In addition to corrosion problems, another serious problem occurs when attempting to run high ionic strength buffers such as 6 M guanidine HCl ('gel filtration') or 1 M NaCl in acidic or basic conditions ('ion exchange') in a

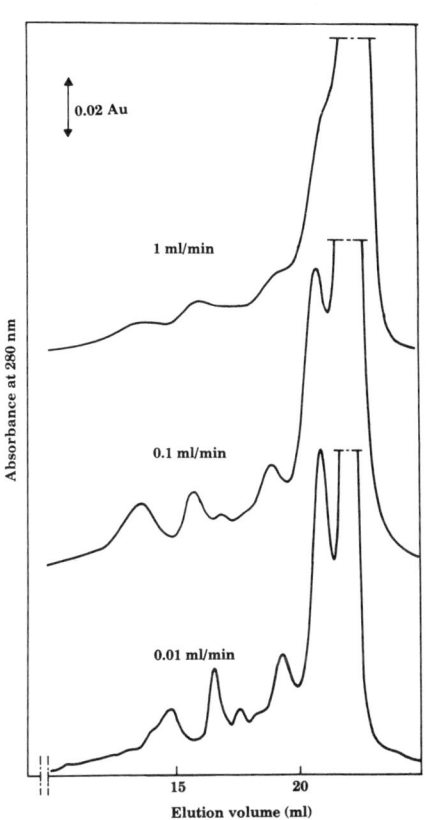

Figure 7. High-speed or high-resolution. A typical result from high performance 'gel filtration'. Sample: human serum. Column: 7.5 × 600 mm LKB 2135 Ultropac TSK G4000SW.

'conventional' HPLC pump. Extremely rapid wearing of the piston seal occurs with subsequent leakage from the pump head. In normal use a small portion of solvent leaks past the piston seal. If the mobile phase contains relatively large amounts of salt, this will crystallize on the piston and then be driven back into the seal as a solid, causing rapid wear of the seal material.

Apart from pump problems, difficulties caused by metal components of the HPLC system are rarely seen if high-quality metals are used in the construction. It is essential that the system is correctly maintained. No part of the system should be allowed to dry out, for example, during use. Buffer salts should always be thoroughly flushed out of the system (with distilled water) before shutting down for extended periods. However, in some 'ion-exchange' applications, metal-ion-chelating to particular ion-exchangers can be a problem, necessitating a fully inert system. The potential danger of deactivating biological samples by metal ions dissolved from metal parts of an HPLC system seems to be a relatively minor problem, considering the absence of comments on this in the scientific literature. The addition of agents such as EDTA or the change of the large surface area metal

components, such as the replacing of solvent filters by non-corroding materials
(e.g., titanium) may be helpful if problems are observed in particular applications.

Complex samples: gradient elution. As stated earlier, the use of gradient elution
is more common in 'biochemical' HPLC because of the frequent requirement to
fully separate complex sample mixtures. For such applications there is a need to
generate accurate, reproducible gradient profiles which can be easily modified.
This is generally achieved by one or other of two principal methods, as illustrated in
Fig. 8: by the use of a switching valve to form eluent concentration gradients on the
inlet (low pressure) side of the pump; or by the use of two pumps driven at different
speeds by a controller, the mixing taking place on the outlet (high pressure) side.

The simple devices normally used for gradient formation in soft gel chromatography are relatively inflexible and have many practical operational difficulties.
The valve-based gradient equipment commonly used for soft gel applications is
not suited for the low total gradient volumes required for the relatively small
columns used in HPLC applications.

Figure 9 outlines the general performance limitations of single- and two-pump
gradient systems. In two-pump systems, the lower flow rate limit of the individual
pumps determines the upper and lower limits of the concentration range that is
attainable at a given total flow rate. This range is decreasing substantially as the
total flow rate is lowered (Fig. 9). This is a second reason for the importance of
low flow rate possibilities, flow optimization for high molecular weight samples
being the first.

For one-pump, valve-based systems, the considerations are different. In order
to cope with small total volume gradients, a small valve-switching cycle volume
is essential, i.e. the cycle of the switching of the valve from buffer A to buffer B

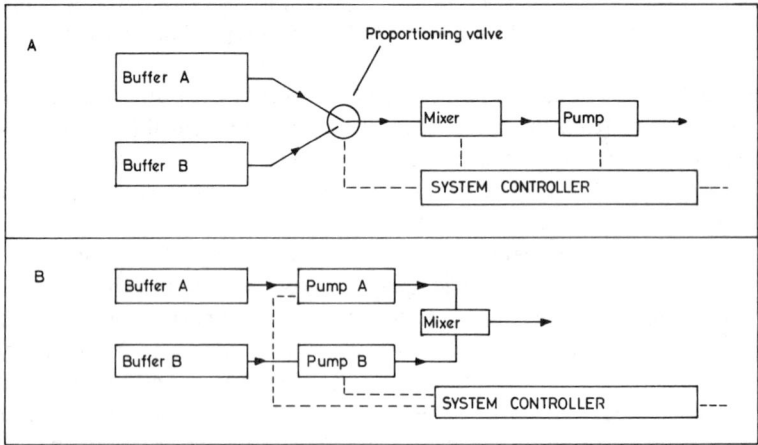

Figure 8. Solvent delivery configurations for gradient formation. (a) one-pump system with
a proportioning valve controlled by a user-preprogrammed controller. (b) A two-pump system
with a controller continuously up-dating the flow rates of each pump, whilst keeping the total
flow rate at user-preprogrammed level(s).

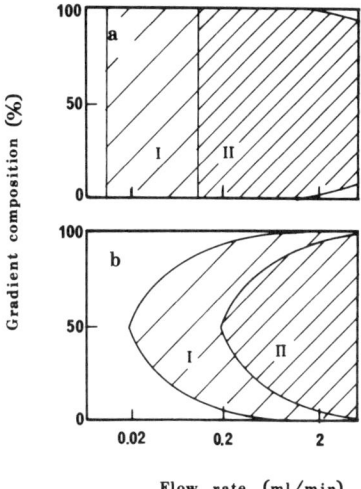

Flow rate (ml/min)

Figure 9. Limitation of gradient systems. The hatched areas illustrate the working ranges for (a) a proportioning-valve-based one-pump system, and (b) a two-pump system. Limiting parameters: minimal flow rate of the pump 0.01 (I) and 0.1 ml/min (II), respectively. A minimal duty time of 25 ms in one valve position and a valve switching cycle volume of 45 μl is assumed for the one-pump system. Reproduced with permission (Sjödahl *et al.*, 1982b).

and back again will generate a volume consisting of plugs of buffer A and B, respectively, which then have to be mixed. At least 100 complete cycles should be generated in a run, implying that gradients from 5 ml total volume and upwards need cycle volumes in the order of 50 μl. A 5% composition thus means a plug of 2.5 μl, equivalent to a duty time of 150 ms at 1 ml/min and necessitating the use of rapid response valves. This explains the decrease in the usable concentration range when increasing the flow rate of a valve-based system, as shown in Fig. 9. However, there is a need for both one-pump and two-pump gradient formation, the choice depending on the fields of application. The advantages of these two modes of gradient formation are, for two-pump gradient formation, (i) the actual composition follows very closely the commands of the system controller as the solvent composition is generated close to the column inlet; (ii) sharp gradient profiles and rapid solvent change-overs can be accomplished as the total internally swept volume is kept to a minimum; and (iii) mixing of the different solvents at high pressure eliminates the need for on-line degassing equipment. The disadvantages are (i) relatively high cost (two pumps); (ii) poor precision of many pumps at low flow rates will limit the accuracy and reproducibility of the gradient at the extremes of composition (<10% and >90% solvent B) even at 'normal' flow rates; and (iii) only a two-liquid (binary) gradient can be formed.

For single-pump gradient systems the advantages are that (i) it can be used with more than two solvents; (ii) lower cost (one pump); (iii) better reproducibility than two-pump systems, particularly if pump specification is limited; and (iv) easier to keep constant flow when composition is changed. The disadvantages are (i) the performance is highly dependent on volumes within the system, pumps, and mixing chambers (gradient profiles will be smoothed); and (ii) if mixing organic and aqueous solvents they must be thoroughly degassed.

The advantages and the importance of the ability to produce *complex* gradients is shown in Fig. 10 which shows optimized 'ion-exchange' HPLC separation of a complex sample of serum from a patient suffering from uremia.

New column technology. There is a high level of interest today in using the newly developed microparticulate stationary phases (5 μm or smaller) packed in smaller diameter columns (2 mm or less, compared with the conventional diameter of 4 or 4.6 mm for analytical chemical applications of HPLC). The background and the implications of this development will be further discussed in a later section of this chapter (Trends in HPLC). As we believe this trend will become important in the biochemical field, some of the implications of this micro-concept on solvent delivery systems are given also below.

In all micro-applications the main consequence of the miniaturization is that separated components elute in very small volumes and that flow rates should be much reduced in order to obtain similar linear velocities to those used in conventional, large-diameter columns packed with the same stationary phase. An analysis of what these chromatographic requirements demand of a solvent delivery system is quite simple. The optimum solvent delivery system should be minaturized with minimal internal volumes and should be capable of achieving low flow rates (down to 10 μl/min) at high working pressures (>300 bar) (Scott and Kucera, 1979). This is the *third* reason for the importance of low flow rate possibilities: flow optimization

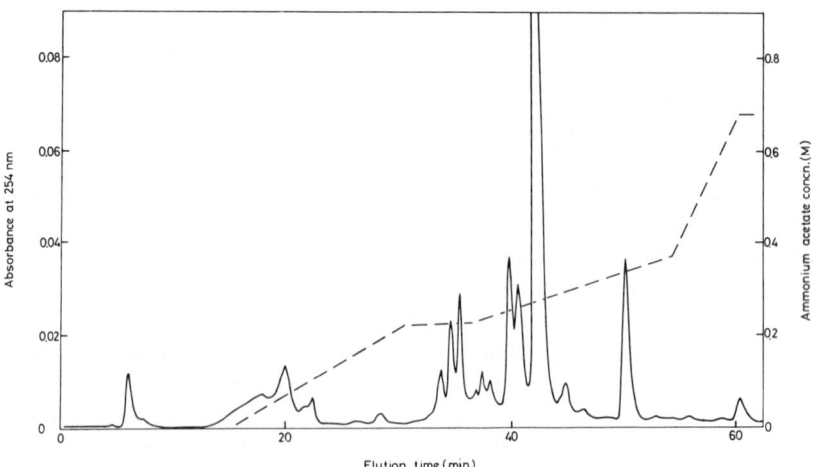

Figure 10. Separation of components which accumulate in body fluids of patients with uremia. The importance of flexible gradient programming. Sample: uremic peptides, mol. wt 350–2000, pooled after gel filtration, lyophilized to 200 μl and then injected. Detection: u.v., 254 nm. Eluent: buffer A 0.03 M ammonium acetate, pH 6.2, buffer B 0.7 M ammonium acetate, pH 6.2. Flow rate: 1 ml/min. Column: LKB 2133 Ultropac TSK 545/DEAE 7.5 × 150 mm. (Reproduced by permission of Dr L. Zimmerman, Department of Renal Research, Huddinge Hospital, Stockholm.)

for high molecular weight samples and good two-pump gradient formation performance being the first and second, respectively.

The other column trend of today will *not* influence solvent delivery systems; the trend of increasing separation speed by decreasing the *length* of the column as well as the bead size of the stationary phase (DiCesare *et al.*, 1981) imply the use of high pressure (>300 bar) *and* high flow rates (5 ml/min).

Total requirements for solvent delivery in 'biochemical' HPLC. In summary, a general-purpose pump for low-molecular weight samples should perform well within the flow rate range 0.01–5 ml/min and should be able to operate against pressures above 300 bars; reflecting a typical 'analytical' requirement. For high-molecular weight samples, especially in preparative applications, the situation is different but the demand similar. Normal-sized (7.5 mm i.d.) analytical and micro-preparative 'gel filtration' columns demand flow rates of 0.01–1.2 ml/min. Larger columns (21.5 mm i.d.) are best utilized in the range 0.1–5 ml/min. For both types of columns the pressures might well be below 5 bars and certainly should not exceed 100 bar. Thus in 'biomacromolecule HPLC', the general purpose pump specification should include flow rates between 0.01 and 5 ml/min and pressures in the range of from <5 to 100 bar.

As a conclusion, the trends of biochemical liquid chromatography, HPLC, and of analytical HPLC, 'micro', require HPLC pumps which are similar to 'conventional' pumps, but with some highly significant differences in specification: flow rate: 0.01–5 ml/min; pressure: 1 to >300 bar; liquid ends: corrosion resistant and organic solvent compatible; and adequate piston seal design, for use even in presence of high salt concentrations. Furthermore, a miniaturized flow path permits one-pump gradient formation of small total volume gradients. The lowering of the minimum flow rate from the pump also expands the range of gradient formation possibilities in a two-pump gradient system.

Pumps. The solvent delivery systems used in conventional soft gel chromatography systems are probably well known to the reader. Gravity flow from a solvent reservoir located higher than the eluate outlet of the system or a peristaltic pump are commonly used in isocratic or step-gradient applications ('gel filtration' and 'affinity' chromatography). 'Ion-exchange' gradient systems use peristaltic pumps often in combination with a gradient programmer to control a low-pressure gradient-mixing valve.

For 'conventional' HPLC applications the most common design of pump has utilized single or dual reciprocating pistons in low volume displacement chambers combined with inlet and outlet check valves (Parris, 1976).

A similar solution, displacement by a linearly moving piston, is used in the syringe type of pump, the major difference being that such pumps are usually used in applications where the volume of one piston stroke is sufficiently large for a complete cycle, thus eliminating the need for automatically reciprocating the piston and for adding inlet and outlet valves.

Both types of pumps exhibit advantages and disadvantages which could be discussed at length. One major difference is that the syringe-type of pump can be easily designed to produce a relatively pulse free flow. The level of pulsation is largely dependent on the stepping motor commonly used to drive the piston. This overall low pulsation level results from the non-reciprocating nature of the piston and the large volume chamber which acts as a pulse dampener. When using the relatively large area pistons needed for adequate solvent capacity, however, high pressure pumps of this type require complex and heavy duty construction due to the enormous forces needed to generate the pressure. Furthermore, the volume of the displacement chamber means that such pumps cannot be used for valve-based single-pump gradient formation. It may thus be concluded, that such a concept of large-area-syringe-type-of-pumps is not suitable for all biochemical HPLC purposes but is very useful for isocratic operations such as 'gel filtration' on today's 10 μm particles or for two-pump gradient formation at medium pressures for short-column or low-flow-rate 'ion-exchange', for example.

A completely different approach, eliminating the problems (for biochemical purposes) of pressure limitations of syringe type of pumps as well as the insufficient working-flow-rate-ranges of 'conventional' HPLC pumps and their non-compatibility with high-ionic strength buffers has recently been described (Sjödahl *et al.*, 1982a, b). This technical solution is reviewed below.

Figure 11 illustrates the design of the liquid end which is responsible for fulfilling the required specification. Compatibility with high ionic strength and corrosive buffers was achieved by lining the pump head with a ceramic displacement chamber, in combination with inert inlet and outlet valves. In order to cope with the requirement for a low flow rate, miniaturized check valves are utilized in twin configurations for both the inlet and outlet valves. Also, the stroke length of the 3-mm diameter piston has been substantially reduced to give a stroke volume of only 36.5 μl (as compared to a typical value of 100 μl for a conventional pump) providing a small total internal volume. The total internal volume of the twin-head pump is less than 300 μl. In addition, this concept also increases the pulsation frequency by a factor of three (flow rate equals stroke volume multiplied by stroke-frequency), facilitating filtering of possible detector noise by electronic means.

A small leakage of solvent past the piston seal is essential for adequate lubrication. However, most commonly used biochemical buffers would precipitate after such leakage. The rinsing port configuration allows for a small, easily washed continuously wetted, piston jacketing space which dissolves precipitates and thereby eliminates the risk of seal or piston damage. Automatic piston and seal flushing is achieved by attachment of a rear membrane to the piston, thus creating an extra membrane pump. Water (or other solvent) is able to continuously flow over the rear of the piston seal and piston whilst the pump is in operation.

Figure 12 summarizes over a period of six days the long term stability, accuracy, and reproducibility of the flow rate that can be achieved in a pump specifically designed to operate at low flow rates. The long term stability at 10 μl/min is clearly very satisfactory. Figure 13 shows the effects of pump design on levels of pulsation in an HPLC system. High flow rate pulsation levels create noise from the detector and limits detection levels that can be achieved. This effect is particularly pronounced with refractive index and electrochemical detectors.

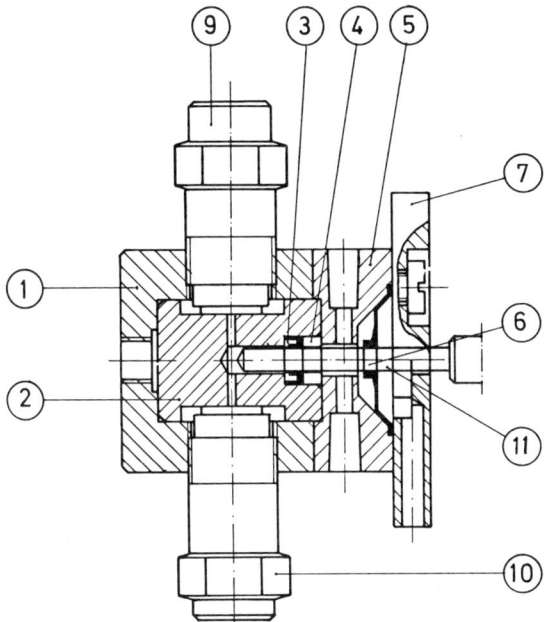

Figure 11. An inert HPLC pump liquid end. The figure illustrates the design of the ceramic lined liquid ends, highlighting the seal rinsing concept which allows 'desalting' of the piston of the LKB 2150 HPLC pump. 1 Pump head housing, 2 displacement chamber (ceramic), 3 piston seal, 4 back-up ring, 5 rinsing port, 6 piston cleaning membrane, 7 drain plate, 9 outlet valve (wetted parts inert), 10 inlet valve (wetted parts inert), 11 piston. (Reproduced by permission of LKB-Produkter AB, Bromma, Sweden.)

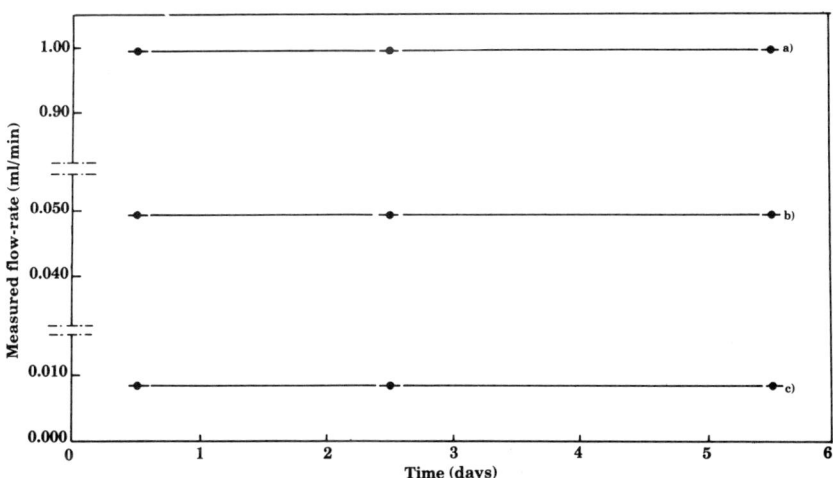

Figure 12. Flow rate stability, reproducibility and accuracy. The figure shows the measured flow rates (● mean value, − limit for ± standard deviation; coincides) over a period of six days. Flow rate settings: (a) 1.00, (b) 0.05, and (c) 0.01 ml/min. The pressure was approximately: (a) 3.5, (b) 0.17, and (c) 0.03 MPa. Pump: LKB 2150 HPLC pump. [Reproduced with permission (Sjödahl *et al.*, 1982b).]

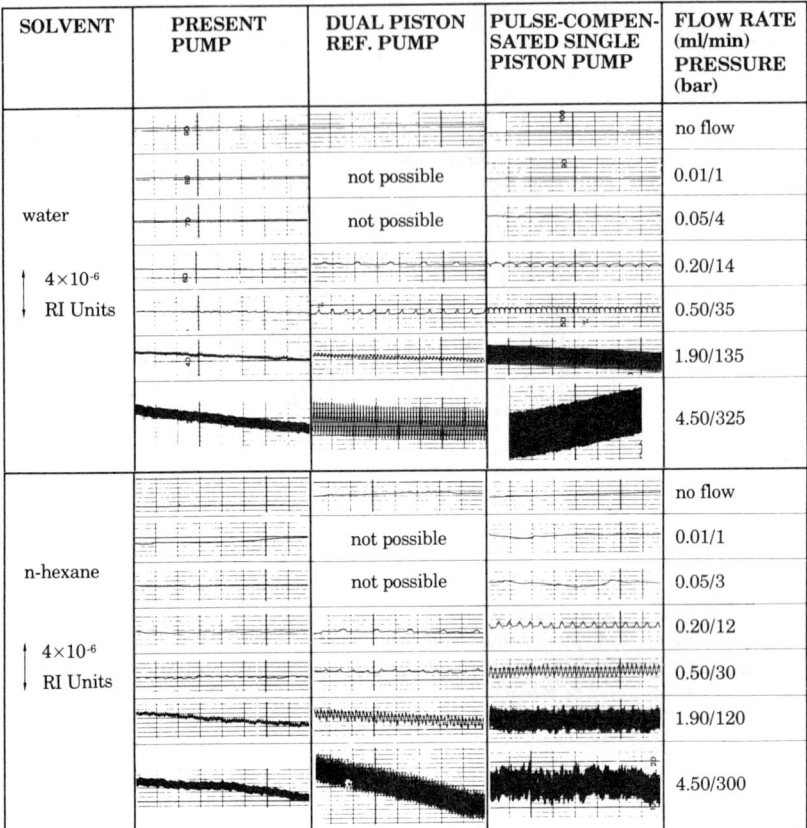

SOLVENT	PRESENT PUMP	DUAL PISTON REF. PUMP	PULSE-COMPEN-SATED SINGLE PISTON PUMP	FLOW RATE (ml/min) PRESSURE (bar)
water 4×10^{-6} RI Units				no flow
		not possible		0.01/1
		not possible		0.05/4
				0.20/14
				0.50/35
				1.90/135
				4.50/325
n-hexane 4×10^{-6} RI Units				no flow
		not possible		0.01/1
		not possible		0.05/3
				0.20/12
				0.50/30
				1.90/120
				4.50/300

Figure 13. Flow pulsation analysis. Comparison of refractive index monitor responses for different pumps under various conditions. 'Present pump': LKB 2150 HPLC-pump. [Reproduced (Sjödahl *et al.*, 1982b).]

The pumps used for comparison to the above described pump were commerically available high-quality 'conventional' HPLC pumps. The dual-piston reference pump was originally equipped with a relatively large (6 ml) pulse dampener, which was removed in order to obtain an adequate comparison. Remounting of this pulse-dampener, of course, improves the situation for the reference pump, but when adding the same dampener to the present pump the same *factor* of improvement is observed.

By analysing these results some interesting conclusions can be drawn:

(a) a specification '0 to x ml/min' is meaningless; some low flow rate limit must be stated even if, admittedly, zero flow rate is possible (dual piston reference pump, Fig. 13);

(b) as expected the dual piston pumps give superior performance compared to the single piston pump examined, when tested under identical conditions;

(c) a pump with a low level of flow rate pulsation can easily increase the sensitivity by a factor of 10 (as shown in Fig. 13, 1.9 ml/min) by decreasing the noise level generated in flow sensitive detectors (or post-column reactors);

(d) the addition of sufficiently large pulse dampeners can, of course, eliminate flow pulsations from any pump. Such systems, however, exhibit large dead volumes which affect the gradient forming capabilities in valve based systems due to the smoothing of the concentration profile, reducing the possibility of running complex, small total volume gradients, and in two-pump systems due to the increased effect of compressibility when the dead volume of the system is increased (Martin and Guiochon, 1978); and

(e) a comparison should always be performed by using an identical pressure transducer — there are huge differences in the inherent response-speeds as well as in the dampening factors introduced by the manufacturer.

Sample injection techniques

There are several reasons for developing methods of sophisticated (as compared to soft gel systems) sample injection in HPLC: (1) at the high pressures commonly employed opening and closing of the column inlet is not practical; (2) through-the-pump sample injection will cause considerable band broadening which is not compatible with the inherent performance of the system, except when the sample can be sharply adsorbed on top of the column; and (3) automating the injection process permits high sample throughput coupled with total automation of the chromatographic system. There are many different types of manual and automated instrumentation for sample injection on the market today, many of them mainly used for routine applications in analytical chemistry. For a review of general technical aspects of automatic injection see Parris (1976).

When we consider biochemical aspects of sample injection there are many points where the ideal specifications are identical to those required by analytical chemists. One major difference between the biochemist and the analytical chemist, however, is the frequent requirement of the biochemist to purify, often on a relatively large scale, amounts of materials that would overload an 'analytical' column. The logical solution of using a larger diameter column may not be practical for economic reasons. In such cases repetitive single-sample injection on an analytical column can be very useful and also, in many cases, attainable without the need for much additional instrumentation.

A basic configuration for automated repetitive injection is shown in Fig. 14. It differs from a conventional manual HPLC system in two ways. Firstly, the addition of a remotely controllable peristaltic pump filling the loop of the injection valve prior to injection. Secondly, the modification/exchange of the manual injection valve into an externally controllable high pressure switching valve. Control of the switching valve and peristaltic pump is performed by the controller used for gradient programming.

Figure 15 illustrates the results obtained from this injection system. The figure indicates the potential of adding these relatively simple components to already existing equipment in situations when (1) the sample is too large to be introduced in a single injection (as exemplified by Fig. 15), or (2) aliquots of the sample is to be analysed on a regular time-course basis in, for example, process control on

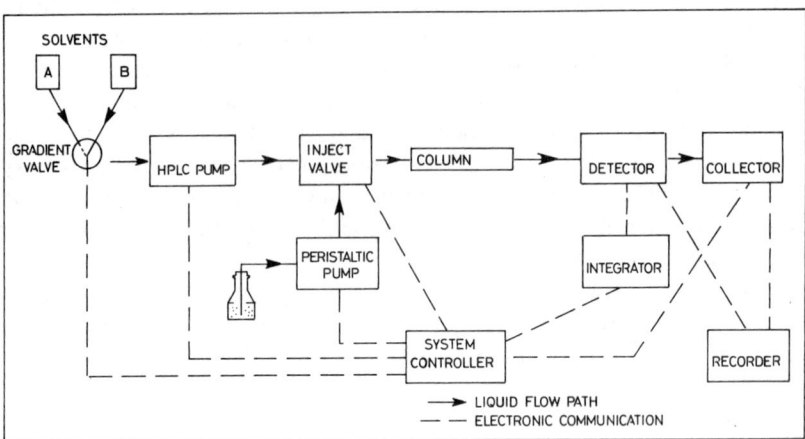

Figure 14. Repetitive single-sample injection – system configuration.

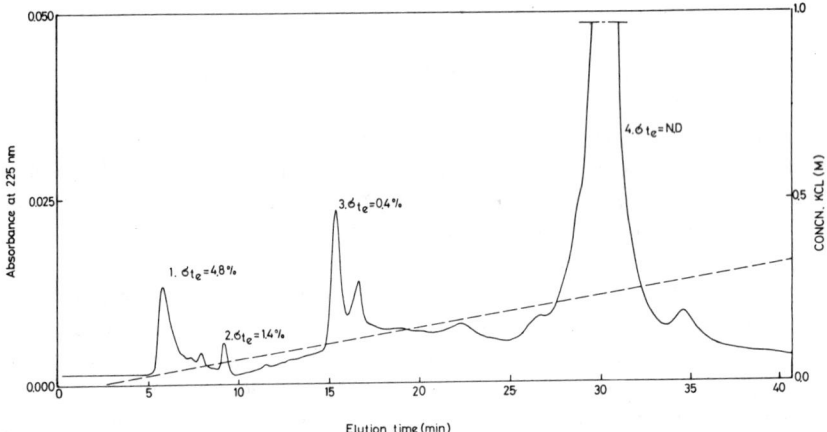

Figure 15. Scaling up by repetitive injection on an 'analytical' column. The figure shows the result from *one* injection out of ten repeatedly performed. σ_{te} is the relative standard deviation of the observed elution times for an individual component. Eluate was collected by automated peak collection to reduce the number of tubes. Sample: mouse serum, diluted 50 times. Detection: u.v., 225 nm, 0.05 AUFS. Gradient: Buffer A 0.05 M Tris-HCl, pH 7.0, buffer B 0.05 M Tris-HCl, 1 M HCl. Flow rate: 1 ml/min. Column: LKB 2135–645 TSK DEAE, 7.5 × 75 mm. Injection volume: 100 μl for each injection.

industrial or laboratory scale. An example of such 'process control' in the ordinary laboratory scale could be the automation of monitoring the progress of a preparative enzymic digestion by, for example, high speed 'gel filtration'. The potential of this method *per se* was indicated by Sjödahl and Winter (1982) and is shown in Fig. 16. These particular results were achieved by manual injection, but the application easily lends itself for automation.

Fraction (or peak) collection
In the past, collection was almost exclusively performed by cutting the eluate

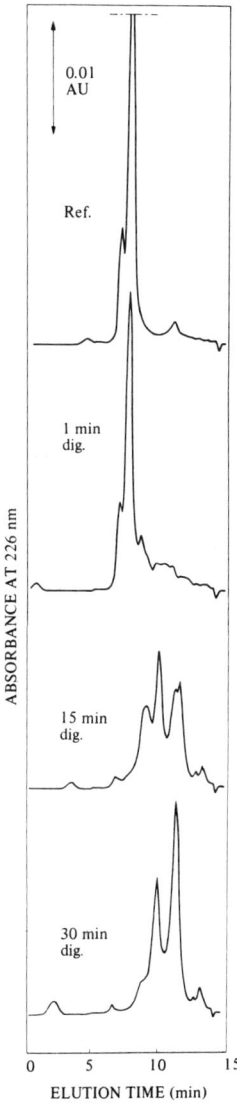

Figure 16. High speed gel filtration, for the monitoring of preparative digestions. Samples: 50 μg aliquots of a tryptic digest of protein A from *S. aureus*. The digestion was performed in 0.1 M Tris-HCl at pH 7.5, 4°C, at an enzyme/substrate ratio of 1:100 (w/w). Columns: LKB UltroPac TSK G2000SW, pre-column (7.5 × 75 mm) plus 7.5 × 300 mm analytical column. Flow rate: 1 ml/min. [Reproduced with permission (Sjödahl and Winter, 1982).]

stream into uniform fractions, collected in test tubes. The main reason for this method of collection was the low availability of on-line detection equipment in the biochemical laboratory: a large number of fractions had to be collected in order not to lose the resolution of the separation system, and then assayed by means of, for example, u.v.-absorptive measurements in a spectrophotometer. For the biochemist it was quite adequate, since several chromatographic systems are run in parallel and the eluates must be collected. Furthermore, the lack of biochemically compatible HPLC columns made sophisticated liquid chromatographic equipment inaccessible to the biochemist. Thus, slow separations in combination with relatively large peak volumes produced this time-consuming and labour-intensive procedure.

However, the situation has changed; much of the work previously carried out manually can be performed by an 'intelligent' fraction collector, peaks are generated when compounds elute from columns connected to on-line detectors, so why not collect peaks? Such a collector would discriminate between interesting peaks and peaks of no interest as well as baseline drift and baseline noise; it would subsequently command the eluate collecting system to collect the material in the final state, eliminating the need for tedious, manual fraction-pooling, and thus also increase the recovery of the purified sample components.

Peak collection is facilitated if the following criteria for instrument design are fulfilled: the peak discriminator is an integral part of the fraction collector to allow combination with any detector, and the algorithm used for peak discrimination is able to cope with drifting and noisy baselines. In this section we will be reviewing Garpe *et al.* (1982) describe peak collection and demonstrate the collection results that can be obtained using a fraction collector with an integrated peak discriminator.

The peak discriminator uses an analogue-to-digital converter and samples the detector signal at pre-programmed intervals. The sampling rate is adjusted to give a frequency corresponding to ten samples across the width of the peak measured at the half-height (see below).

Four parameters are used by the peak discriminator: threshold (TH), peak width at the half-height (PW), isocratic point (IP), and asymmetry factor (AF). Threshold and peak width must always be set; the other parameters are optional for use in special chromatographic applications. The threshold (TH) sets an absolute minimum to the height above the baseline of an accepted peak. This parameter can be compared to a sophisticated level sensor. The peak width (PW) sets the sampling rate of the discriminator (see above) as well as, in combination with TH, the slope level criteria for peak discrimination. Slopes larger than the maximal slope of a gaussian peak, with peak height and peak width at half the height equalling TH and PW, respectively, will be considered as a potential peak if the criteria are fulfilled for two successive samples, together with a positive slope for three samples. In contrast to the older method of 'level sensing' or 'peak cutting' where peaks are rejected only if the baseline plus the peak signal keeps below a preset value, this method is largely independent of the effects of baseline drift and/or noise.

Figure 17 shows the ability to discriminate between peaks of adequate size, peaks which are too small, noisy baselines (or air-bubble spikes), and drifting baselines. The figure also highlights the difference between this two-dimensional approach and that of the one-dimensional level-sensing or peak cutting method. By combining various TH and PW values, it is thus possible to optimize parameters for peak detection in any particular chromatographic situation.

For peak exit, several alternatives are considered, to allow for the clear

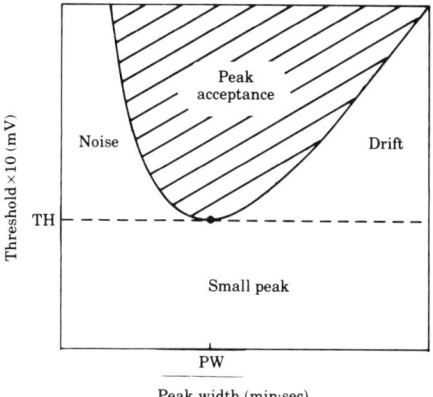

Peak width (min:sec)

Figure 17. Peak collection: criteria for rejection/acceptance. Peak discriminating algorithm in the LKB 2211 SuperRac fraction collector. TH and PW represent the programmed values of threshold and peak width, respectively. The threshold is expressed in signal voltage, and the peak width in elution time. The non-shadowed areas represent situations where peaks are rejected. [Reproduced with permission (Garpe *et al.*, 1982).]

identification of, for example, plateaus and non-resolved peaks. Furthermore, the peak entrance criterion is lower during a time of PW following peak exit, facilitating a more rapid response to a subsequent non-resolved peak. The optional parameter isocratic point (IP) defines the retention time chosen for peak-width setting, and is used in isocratically eluted systems to compensate for peak-broadening as a function of time.

The second optional parameter, asymmetry factor (AF), allows peak-entrance and peak-exit criteria to be fulfilled at different slopes and, thus, compensates for leading or tailing phenomena observed in some column liquid chromatography systems.

To take advantage of the intelligent peak discriminator requires correct design of the fluidic system. If a switching valve is not used the situation is relatively simple: peaks are collected in one fraction size (or in a single tube) – non-peaks (eluate giving baseline signal) are collected in a larger fraction size. Of course, a delay function compensating for the volume between the detector cell and the flow nozzle should be utilized.

The introduction of a switching valve between the detector and the collecting vessels adds a new dimension. From the user's point of view, the most severe effect of an improper design would be the possible carry-over effect, resulting in contamination (or re-mixing) of the separated components. In order to minimize such effects, tube change and valve switch triggering can be induced differently by an identified peak (Fig. 18). An ideal detector with a perfectly straight baseline eliminates the need for sophisticated peak sensors. This situation is, however, not encountered very often. Baseline drift may originate from the detector itself or may be due to alterations of the eluant composition in gradient applications. The situation is particularly severe in slow or in repetitive high-speed applications. High sensitivity analysis with a great deal of noise does not improve the situation.

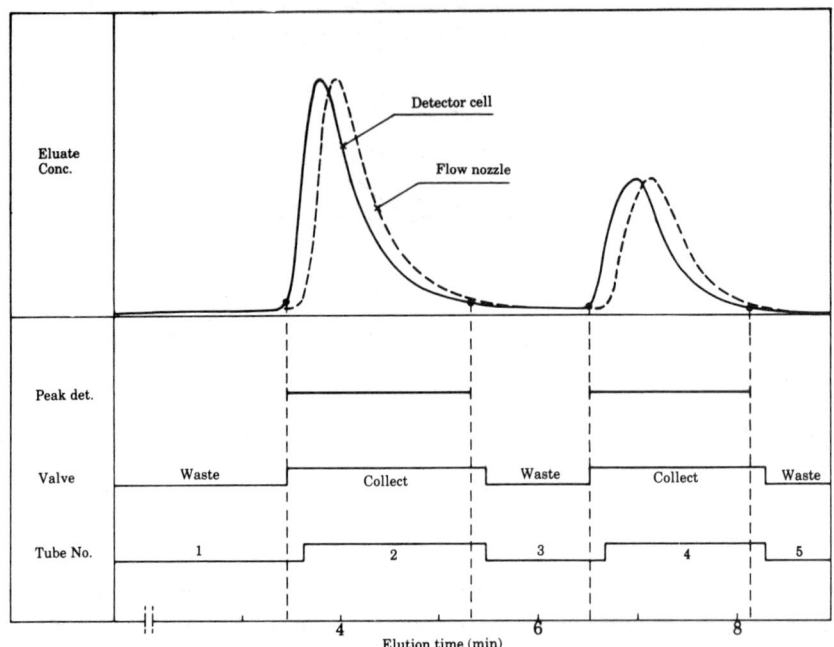

Figure 18. Fluidic-system response in automatic peak collecting mode. The relationship between valve switching and tube change in the LKB 2211 SuperRac fraction collector are shown. Sample: 100 μl of vitamin B_{12} and phenol, 2 mg/ml of each. Column: 4.6 × 260 mm LiChrosorb RP 18, 10 μm homepacked. Flow rate: 2 ml/min. Elution: methanol, 30% (v/v). Peak-discriminator parameter settings: PW = 15 s, TH = 0.5 mV, AF = 3.0. The horizontal bars below the peaks illustrate the valve position and the tube change as functions of the entrance and exit of peaks (solid dots). [Reproduced with permission (Garpe *et al.*, 1982).]

Figure 19 shows an extremely noisy chromatogram at high sensitivity. In this application the time window programs of the fraction collector were used to define interesting areas in the chromatogram. Outside these windows the eluate was diverted to waste. Collection of different fraction sizes is an alternative. The result is surprisingly good: hundreds of non-peaks are rejected and only the peaks of interest are detected/collected. In Fig. 20 a highly drifting baseline was generated in a short-wavelength u.v.-detected gradient application, however, this did not adversely affect the quality of the peak collection. Figure 21 shows a typical biochemical chromatogram: high resolution 'gel filtration'. Such applications are characterized by relatively slowly eluting peaks and call for wide peak-width setting possibilities.

If considering the end-result of an HPLC purification, the tubes, it is interesting to compare the result illustrated in Fig. 21 to what could be expected from a separation where the eluate was collected by means of conventional fraction collection. Many more tubes would have resulted — some 80 tubes as compared to seven 'peak tubes' (and seven 'valley tubes') in the automated mode.

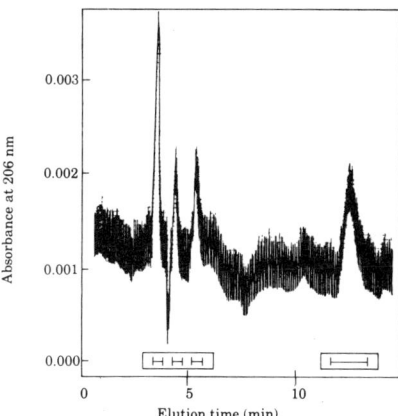

Figure 19. Peak discrimination in high-sensitivity application. Sample: 20 μl of a mixture of dimethylphthalate, diethylphthalate, and dibutylphthalate, 6–20 ng of each. Detection: u.v., 254 nm. Column: 4.6 × 250 mm LiChrosorb RP 18, 10 μm, homepacked. Flow rate: 1 ml/min. Elution: methanol 80% (v/v). Fraction collector: LKB 2211 SuperRac. Peak-discriminator parameter settings: PW = 20 s, TH = 3 mV, IP = 4 min, AF = 1.0, window 1: 3–6 min, window 2: 11–13 min. Detector time-constant: 0.5 s. The horizontal bars below the peaks represent peak identification; the surrounding rectangles represent the set time windows. [Reproduced with permission (Garpe *et al.*, 1982).]

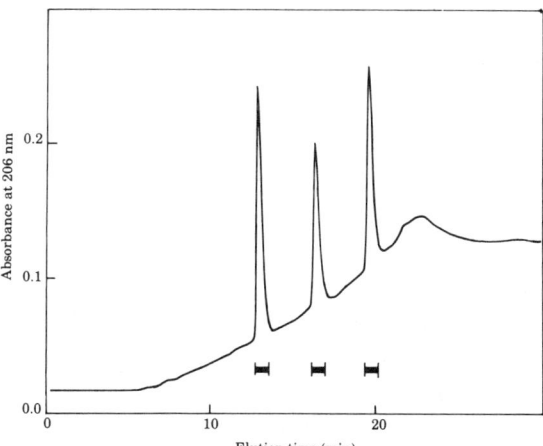

Figure 20. Peak discrimination in a gradient application. Sample: 20 μl of a mixture of methyl-4-hydroxybenzoate, ethyl-4-hydroxybenzoate, and propyl-4-hydroxybenzoate, 0.2 mg/ml of each. Column: 4.6 × 250 mm LiChrosorb RP 18, 10 μm homepacked. Flow rate: 2 ml/min. Elution: solvent A: methanol, 20% (v/v). Solvent B: methanol, 65% (v/v). Solvent A for 2 min, linear gradient to solvent B for 18 min, then solvent B. Fraction collector: LKB 2211 SuperRac. Peak discriminator settings: PW = 10 s, TH = 2 mV, AF = 2.0. The horizontal bars below the peaks represent peak identification. [Reproduced with permission (Garpe *et al.*, 1982).]

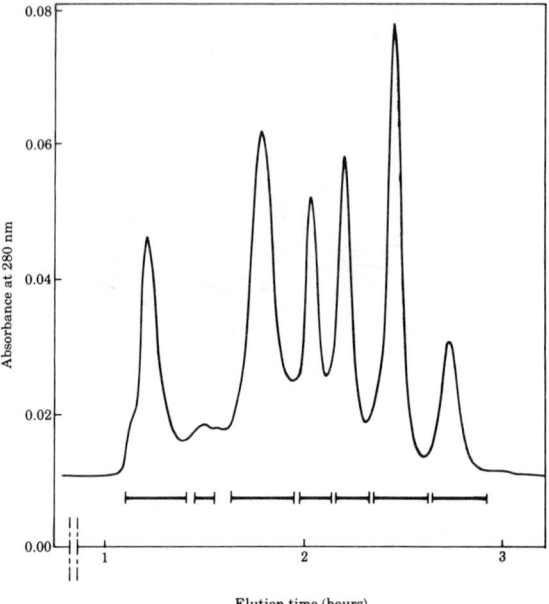

Figure 21. Peak discrimination in high-resolution 'gel filtration' of proteins. Sample: 200 µl of a mixture of fibrinogen, γ-globulin, human serum albumin, and myoglobin, 1–3 mg/ml of each. Column: LKB 2135 UltroPac TSK G3000SW, 7.5 × 600 mm. Flow rate: 130 µl/min. Elution: 0.1 M Na_2HPO_4: NaH_2PO_4, 0.1 M NaCl, 0.05% (w/v) NaN_3, pH 6.5. Fraction collector: LKB 2211 SuperRac. Peak-discriminator parameter settings: PW = 2 min 25 s, TH = 0.5 mV, AF = 1.0. The horizontal bars below the peaks represent peak identification. The last peak represents low molecular weight impurities of the sample proteins.

TRENDS IN HPLC

New column concepts and new packing materials continue to be developed. This in turn places further demands on chromatographic instrumentation and eventually and hopefully will lead to the availability of instrumental systems optimized for the state-of-the-art columns available at a specific time, i.e. the next optimized system might be VHPLC (very high performance liquid chromatography). The question is then; how to abbreviate the subsequently appearing systems? Today, it is possible to clearly identify two major trends which are of general interest to all chromatographers, biochemists included: the concept of 'micro-HPLC' and multi-channel detection by linear photodiode-array (PDA) devices. A brief survey of these new techniques and the possibilities they offer, seen from a biochemist's point of view, will be presented below.

The micro concept

The concept of micro-chromatography can be separated into two main areas. The first of these offers significant improvement in the time needed for separation.

Rapid HPLC. The availability of even smaller spherical particles (3 μm) than those commonly used in HPLC today (5–10 μm) has led to them being used in relatively short columns at linear flow rates generally higher than those used with conventional HPLC media (DiCesare *et al.*, 1981). The major advantage of this is to make the already fast analysis even faster. The technique is sometimes referred to as 'very-high-speed-HPLC' but with present rates of development this nomenclature may be obsolete in some years' time. At the time of writing no stationary phases of this type are available for the separation of biomacromolecules. For low-molecular weight biomolecules, however, the 'reversed phase' mode can be a powerful tool.

In all probability such methods will be adapted for separations of biomolecules when suitable packing materials are available. The present state of the art is shown in Fig. 22, which shows the result of performing 'gel filtration' of a mixture of proteins on a short column at a high linear velocity. The packing material used in this example is 10 μm; run times of less than 1 min should be possible with smaller diameter packing materials.

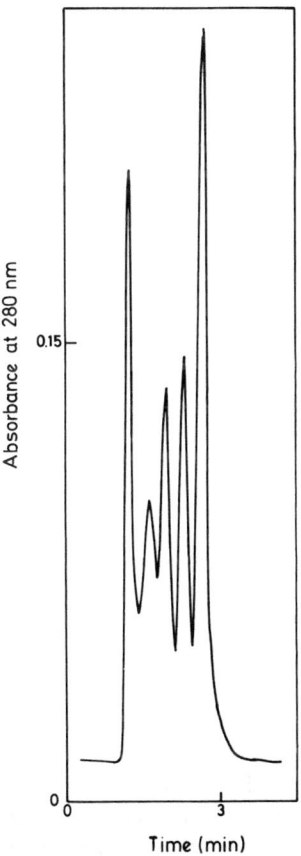

Time (min)

Figure 22. 'Very high speed gel filtration'. Sample: mixture of thyroglobulin, γ-globulin, ovalbumin, myoglobin, and vitamin B_{12}. Detection: u.v., 280 nm. Eluent: 0.1 M sodium phosphate, pH 6.8. Flow rate: 1 ml/min. Column: LKB 2135–075 TSK GWSP, 7.5 × 75 mm.

The expected improvement of 'biomacromolecular' separation media towards even smaller particles as well as the tremendous possibilities already available through '3 μm-reversed phase chromatography' thus put even greater demands on the instrumentation. The requirements for solvent delivery (mainly flow rate/ pressure considerations) were discussed in a preceding section. For the other components of the chromatographic system, factors such as capillary lengths and diameters, flow-cell volume, detector time constant, switching valve response times, and tube change times for fraction collectors must be optimized if high speed is to be combined with high resolution and reliability.

Microbore HPLC. This concept of HPLC is concerned with instrumental aspects of internal volumes, solvent flow rate, and system geometry. In principle, it has little or nothing to do with column packing materials. Here columns with a smaller (typically 1 mm) internal diameter are packed with 10 μm particles. The first high efficiency (long) microbore columns were described by Scott and Kucera in 1979. An earlier description of the micro concept can be attributed to Kuzmin *et al.* (1969) who actually also included multichannel u.v.-detection; the other major current instrumental trend. The amino acid analyser was one of the first (again) commercially available chromatographic systems to use a small (relative to the standard at the time of introduction) diameter analytical column.

Some confusion exists in the literature as to the practical benefits to be derived from microbore HPLC. Two major advantages are: (1) the elution of a given volume of sample in a decreased elution volume, resulting in an increased mass sensitivity, and (2) a reduced volumetric flow rate, given a specific linear flow rate or a specific analysis time, resulting in a decrease (proportional to the square of the reduction of column diameter) in solvent consumption and consequently increasing the possibility of interfacing the separation system with mass spectrometric detection (Alcock *et al.*, 1982). One possible minor advantage is the possibility of increasing separation efficiency by connecting columns in series. A common opinion, not sufficiently well documented, is that, due to the miniaturization of column size, column efficiency comes closer to theoretical expectations. It is important to note that 'microbore HPLC' is no faster than 'conventional' HPLC.

The advantages of low solvent consumption are clear enough. Running costs for the system can be greatly reduced: a 4.6-mm column consumes more than 20 times the amount of solvent consumed by a 1-mm column. The low flow rates used may permit the use of constituents in the mobile phase so expensive that they would be prohibitive in a conventional system. In addition disposal of the used solvents, which can present a major problem and expense, is on a correspondingly smaller scale.

The increase in mass sensitivity from a microbore HPLC system is dependent not only on the geometry of the column but also on the injection volume, which must be decreased in direct proportion to the square of the column diameter. A 1-mm column exhibits a typical optimal injection volume of around 1 μl.

If the sample can be adsorbed directly on top of the column prior to elution, it is very easy to avoid the common sample waste often accompanying the injection of very small sample volumes just by diluting the sample to a practical volume. If this is not the case, one often has to accept to inject only a portion of the sample onto the micro-column. This drawback is, in our opinion, however, a *very* important advantage in the vast number of biochemical applications, for which the amount of sample injected onto the HPLC column represents wastage: the residual amount of sample is vital and it has to be used for further studies.

To maintain separation efficiency with the smaller volume peaks eluting from a microbore column, it is, of course, necessary to decrease the internal volume of the detector cell. However, care must be taken that this is not achieved at the expense of the flow cell pathlength (if using u.v. detection) and hence the detector signal which will determine the signal-to-noise ratio normally referred to as a measure of sensitivity. Many micro-flow cells have a path length of only 1 mm. This largely counteracts one of the major advantages of the concept, the increased mass sensitivity has to be met by an adequate cell design when concentration sensitive detectors (like the u.v.-detector) are used in micro-systems. A microbore HPLC system, equipped with an optimized low volume flow cell (0.8 μl, length 3 mm) in our experience easily gives a higher than five-fold increase in sensitivity when compared to a conventional system. A discussion on this most debated benefit of miniaturizing the chromatographic system was published by Bowermaster and McNair (1983).

Although microbore HPLC has been utilized to date mainly by analytical chemists, the system offers further potential advantages to the biochemist. In situations where sample amount is limited (providing the sample can be concentrated), microbore methods can be invaluable exemplified by the fact that modern methods for protein microsequencing place higher demands on methods for PTH amino acid determination. Using a microbore system for this application, as shown in Fig. 23, detection limits can be decreased significiantly. Analysis of PTH-amino acids probably, in our opinion, represents the first major biochemical application for microbore HPLC and many more are to be expected in the future.

Diode array u.v.-detection
Recent developments in technology have led to the availability of integrated circuits composed of a linear array of photo-sensitive diodes; the photo-diode-arrays (PDA). Typically, up to 256 light sensitive elements are arranged in a total distance of a few millimetres. When incorporated into a suitable optical system containing a u.v. light source, flow-cell and a *subsequent* monochromator, the PDA can be used for rapid capture of complete u.v. spectra from sample molecules passing through the flow-cell as shown in Fig. 24. Observe that spectral differentiation occurs following the flow-cell.

The combination of the optical unit with a data processing unit (Fig. 24) enables complete information of the u.v.-absorbance of separated components to be obtained, without interfering in any way with the chromatography. Thus u.v.-detected chromatography no longer results in two-dimensional information (absorbance vs time) but can now be three-dimensional (absorbance, time, and wavelength).

The technique is currently being widely adopted within analytical chemistry, with major applications for peak purity analysis and positive identification. However, the larger amount of information available from such a detector means that

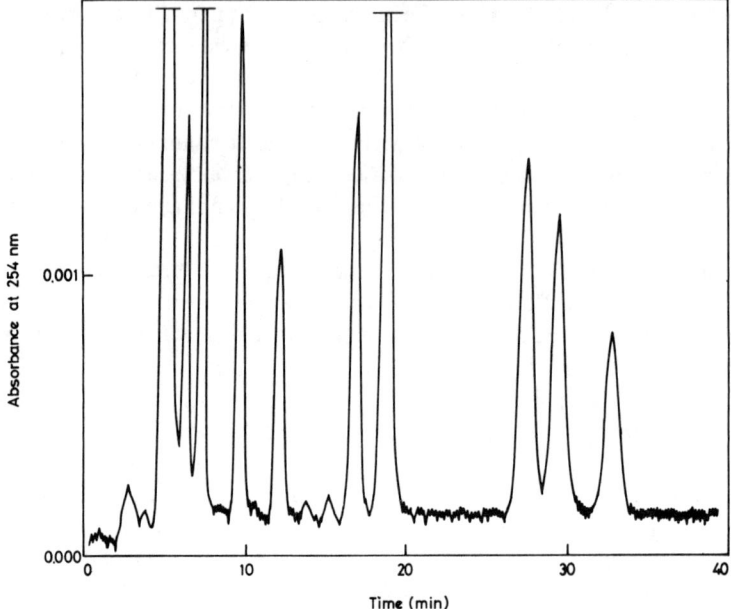

Figure 23. Micro-column separation of PTH-amino acids. Sample: PTH-amino acids, 1–5 pmol each. Detection: u.v., 0.005 AUFS. Eluent: CH_3CN/THF aqueous sodium acetate pH 5.8. Flow rate: 50 μl/min. Column: LKB 2134–600, RP-18 (10 μm), 1 × 250 mm.

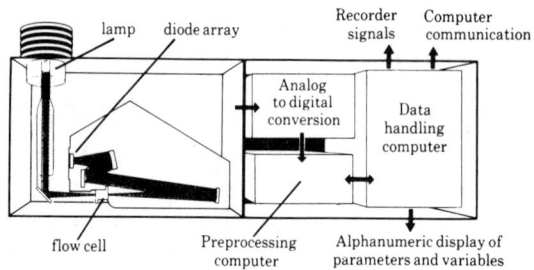

Figure 24. Linear photo-diode-array u.v. detector.

it most probably will rapidly become essential also for the biochemists. Three-dimensional peptide mapping is just one of the exciting possibilities that spring to mind.

For practical use, it is, however, very important to incorporate powerful data processing units capable of handling the huge amount of data generated. By thinking of the detector as a column, i.e. a black box from which a flow of continuously changing information elutes, it is easily realized that a 'fraction collector' might help to take care of the increased amount of information. The combination with a powerful 16-bit desk computer seems to be an adequate choice for the purpose of collecting the rapidly 'eluting' spectra. The concept of spectrum collection is shown in Fig. 25 which also focuses on the computer memory saving concept introduced by the 'window' setting possibilities also used for the same purpose ('memory' saving) in conventional fraction collection: absorbance range, spectrum recording time, and wavelength range can all be 'windowed' (selection of a segment within a range).

During the run the *continuously* growing three-dimensional 'landscape' is rapidly segmented into conventional spectra and displayed as shown in Fig. 26. An alternative is to reduce the data into a reduced 3D-picture: a conventional map called a 'contour plot' shown in Fig. 27. This form of presentation is typical after a completed run and serves as a start for the post-run data analysis.

The advantage of post-run data treatment is self evident, the decision is taken when the overall results are known. By analysing the complete, or reduced, 3D-picture, it is now possible to convert the data to a conventional form by 'slicing' in the wavelength domain (selecting a wavelength for presentation of a conventional chromatogram) as well as in the time domain (selecting a retention time for presentation of a particular spectrum). Different results are achieved by altering the width of the slices, as shown in Figs 28 and 29 showing the benefits of increasing the wavelength band-width to decrease selectivity and increase sensitivity and of varying the signal-averaging time to optimize sensitivity, respectively.

Additionally, in principle, any arithmetic or algebraic treatment of the signal is possible. This is illustrated in Figs 30 and 31 illustrating the use of differential absorbance u.v.-detection to enhance selectivity and of absorbance ratioing to analyse peak purity, respectively. A pure peak is, of course, expected to exhibit a ratio constant over its full width.

To summarize, u.v.-absorbance measurements by the use of linear PDA-detectors offer a very powerful tool to fully benefit from all information inherently available

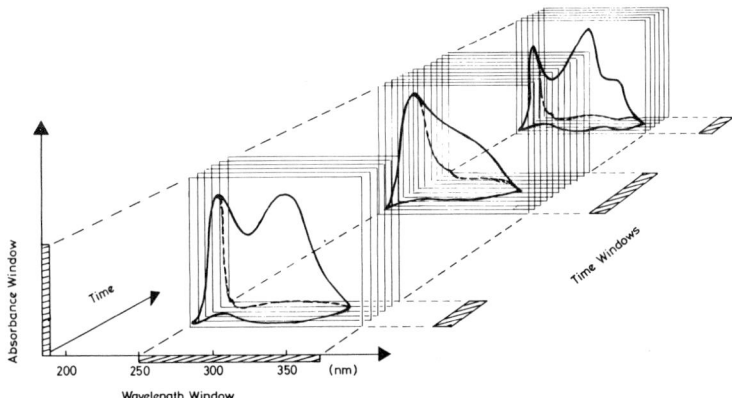

Figure 25. The concept of spectrum collection. The figure shows the similarity of data collection from a u.v.-photo-diode-array-detector to that of fraction collection. By combining the LKB 2140 Rapid Spectral Detector with an IBM Personal Computer it is possible to collect spectra at different times (indicated by the thin *x–y* planes) by 'windowing' and, thus, focusing on regions of expected interest. The coarse lines illustrate the 'windowed' parts of a hypothetical elution profile. This is a *continuous* area in the 3D space: commonly represented as spectra subsequent to segmentation.

J. Sjödahl et al.

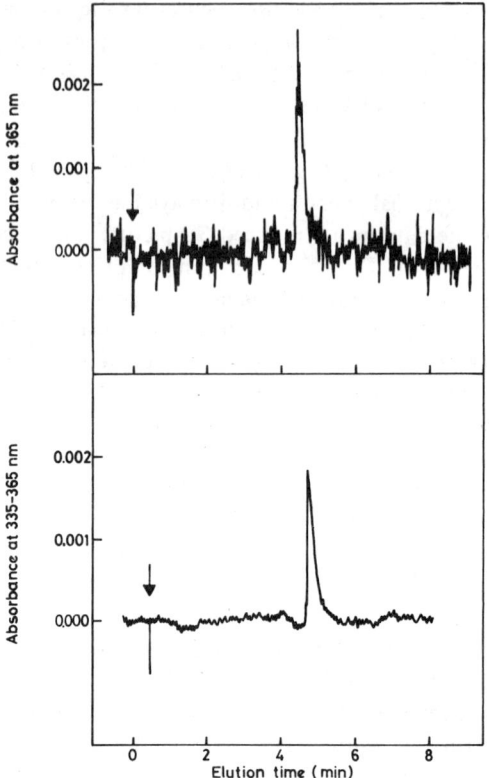

Figure 28. Wavelength 'bunching' for increased signal-to-noise ratio. Detector: LKB 2140 Rapid Spectral Detector. Sample: 80 ng mitomycin. Eluent: 0.01 M Na_2HPO_4, Na_2HPO_4, pH 7.0 in 40% methanol. Column: LKB 2134–215, RP-18 (5 μm), 4 × 250 mm. Flow rate: 1 ml/min.

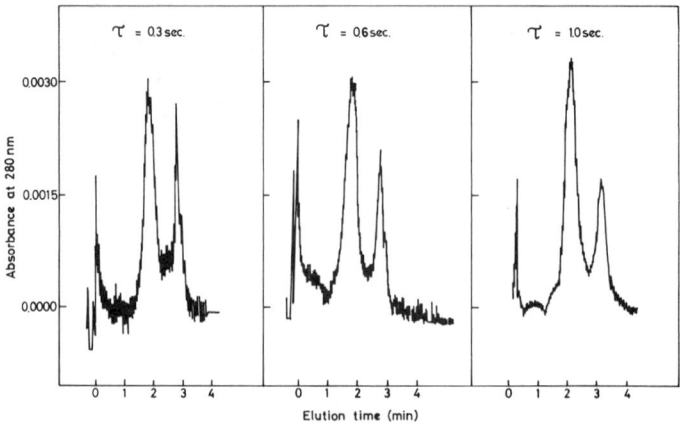

Figure 29. Time 'bunching' for increased signal-to-noise ratio. Sample: 5 μg immunoglobulin G. Detection: LKB 2140 Rapid Spectral Detector, 280 nm. Eluent: 0.01 M NaH_2PO_4, Na_2HPO_4, 0.1 M NaCl, pH 7.0 in 40% methanol. Column: LKB 2135–075 TSK GWSP, 7.5 × 75 mm. Flow rate: 1 ml/min.

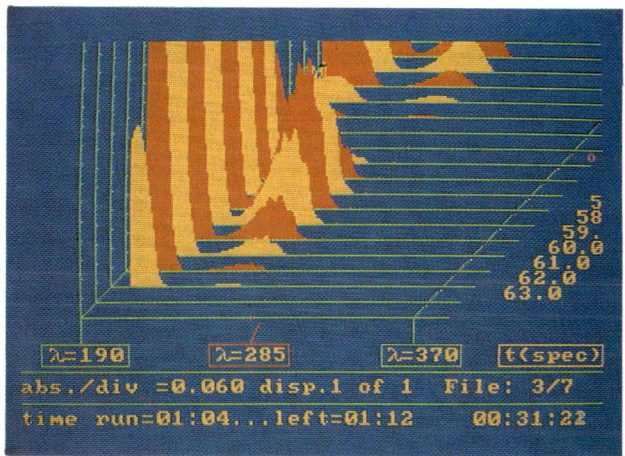

Figure 26. Three-dimensional real-time display from a linear photo-diode-array u.v. detector. Detector: LKB 2140 Rapid Spectral Detector. Display: IBM Personal Computer. Sample: toluene, benzene, naphthalene, and anthracene. Eluent: 80% (v/v) methanol. Flow rate: 3 ml/min. Column: LKB 2134–210, RP-18 (10 μm), 4 × 250 mm.

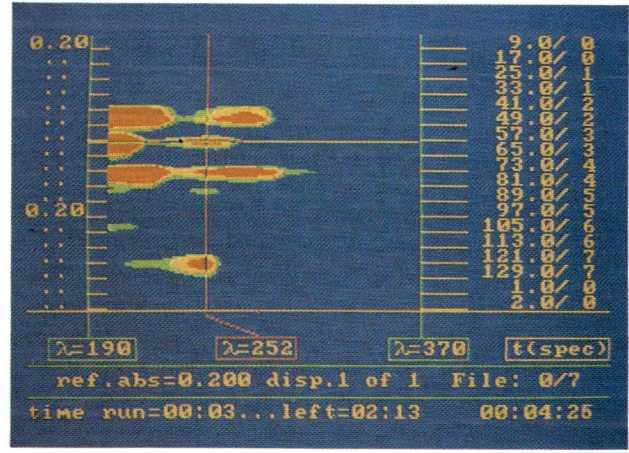

Figure 27. The contour map: the reduced three-dimensional display. Sample: toluene, benzene, naphthalene, and anthracene. Eluent: 80% (v/v) methanol. Flow rate: 3 ml/min. Column: LKB 2134–210, RP-18 (10 μm), 4 × 250 mm. Detector: LKB 2140 Rapid Spectral Detector. Display: IBM Personal Computer.

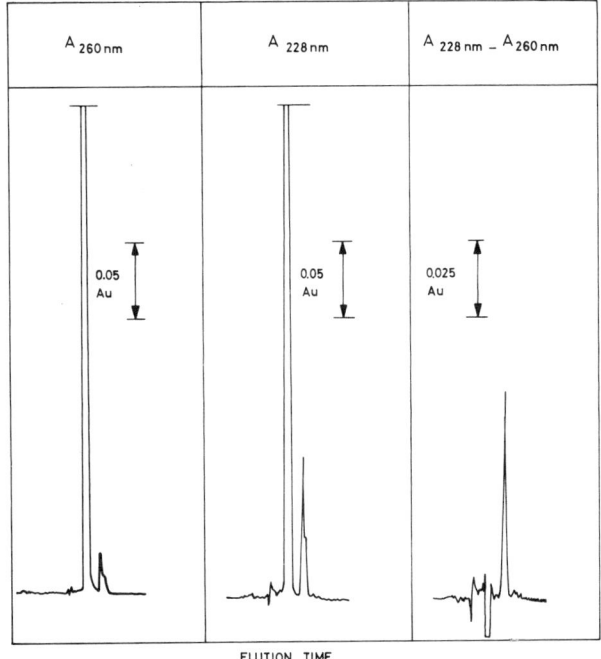

ELUTION TIME

Figure 30. Differential absorbance u.v.-detection. Detector: LKB 2140 Rapid Spectral Detector. Sample: Anthracene 95 ng, naphthalene 5 μg. Eluent: 100% (v/v) methanol. Flow rate: 1 ml/min. Column: LKB 2134–215, RP-18 (5 μm), 4 × 250 mm.

in such applications. The method is expected to be increasingly used also in biochemical applications; for reviews on analytical chemistry applications, see, for example, Fell (1981) and Borman (1983).

CONCLUSIONS

It is possible to identify four main recent developments of HPLC today which will more or less revolutionize the way a biochemist performs a separation by column liquid chromatography. These are:

1. The development of rigid *and* biocompatible high performance packing materials offering the biochemist the key to success in solving liquid chromatography separation problems: small particles useful for the full spectrum of biochemically interesting liquid chromatographic methods.

2. The development of biochemically optimized equipment coping with all types of currently available liquid chromatography separation media.

3. The development of the production of and the packing technology for very small packing materials, further refining HPLC methodology of today.

4. The development of increased selectivity in detection by, for example, linear photo-diode array detectors allowing for simultaneous multi-channel optical detection.

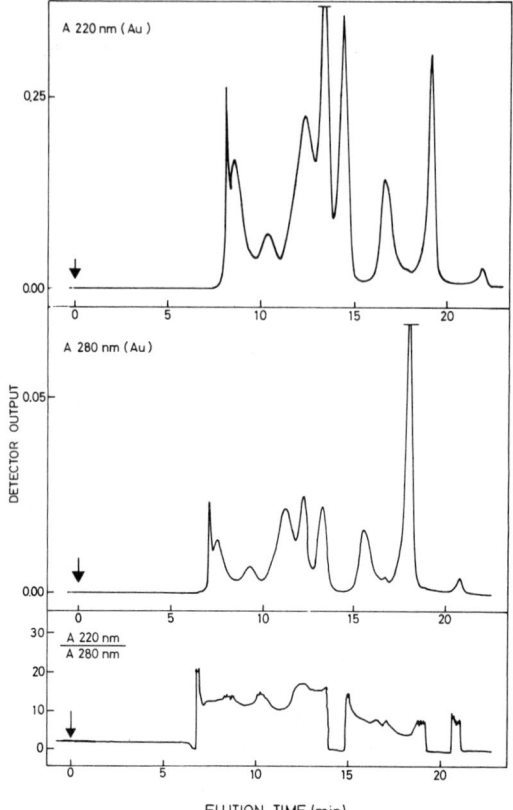

Figure 31. u.v.-Absorbance ratioing detection for peak purity control. Detector: LKB 2140 Rapid Spectral Detector. The potential of detecting impurities in 'pure' peaks is shown. A pure peak should exhibit a constant absorbance ratio through its full width. Sample: thyroglobulin, immunoglobulin G, ovalbumin, cytochrome C, myoglobin, vitamin B_{12}; 0.6 mg total sample amount. Eluent: 0.1 M sodium phosphate, pH 6.5. Flow rate: 1.2 ml/min. Column: LKB 2135–360 TSK G3000SW, 7.5 × 600 mm.

In this chapter, we have tried to describe and analyse these developments and also to give examples of how we believe it is possible to perform state-of-the-art biochemical HPLC by combining existing knowledge within biochemical liquid chromatography and analytical chemical HPLC.

REFERENCES

Alcock, N. J., Eckers, C., Games, D. E., Games, M. P. L., Lant, M. S., McDowall, M. A., Rossiter, M., Smith, R. W., Westwood, S. A., and Hee-Yen Wong (1982). High performance liquid chromatography – mass spectrometry with transport interfaces, *J. Chromatogr. 251,* 165–174.

Borman, S. A. (1983). Photodiode array detectors for LC. *Anal. Chem. 55,* 836A–842A.

Bowermaster, J. and McNair, H. (1983). Sensitivity in microbore HPLC. *Liquid Chromatogr. 1,* 362–364.

Chang, S. H., Gooding, K. M., and Regnier, F. E. (1976). High perfromance liquid chromatography of proteins. *J. Chromatogr. 125*, 103–114.

DiCesare, J. L., Dong, M. W., and Ettre, L. S. (1981). Very high speed liquid column chromatography – the system and selected applications. *Chromatographia 15*, 257–268.

Deyl, Z. (Ed.), Everaerts, F. M., Prusik, Z., and Svendsen, P. J. (Co-eds) (1979). Electrophoresis, a survey of techniques and applications. Part A: techniques. *J. Chromatography Library 18.* Elsevier, Amsterdam.

Ettre, L. S. (1971). The development of chromatography. *Anal. Chem. 43*, 20A–31A.

Fell, A. F. (1981). New dimensions in spectroscopy. *Medical Laboratory World*, Oct., 19–25.

Garpe, L., Lundin, H., and Sjödahl, J. (1982). Intelligent fraction collection. *Sci. Tools 29*, 11–14.

Giddings, J. C. (1965). *Dynamics of Chromatography, Part I.* Marcel Dekker, New York.

Kuzmin, S. V., Matveev, V. V., Pressman, E. K., and Sandokhchiev, L. S. (1969). A simple procedure of quantitative chromatographic analysis on the ultramicroscale. *Biokhimiya 34*, 706–711.

Martin, A. J. P. and Synge, R. L. M. (1941). A new form of chromatogram employing two liquid phases. *J. Biochem. 35*, 1358–1366.

Martin, M. and Guiochon, G. (1978). Theoretical study of the gradient elution profiles obtained with syringe-type pumps in liquid chromatography. *J. Chromatogr. 151*, 267–289.

Moore, S. and Stein, W. H. (1948). Partition chromatography of amino acids on starch. *Ann. N.Y. Acad. Sci. 49*, 265–270.

Nota, G., Marrino, G., Buonocore, V., and Ballio, A. (1970). Liquid-solid chromatography with open glass capillary columns separation of 1-dimethylaminonaphthalene-5 sulphonyl amino acids. *J. Chromatogr. 46*, 103–106.

Parris, N. A. (1976). Instrumental liquid chromatography. *J. Chromatography Library 5*, Elsevier, Amsterdam.

Peterson, E. A. and Sober, H. A. (1956). Chromatography of proteins: I. Cellulose ion-exchange adsorbents. *J. Am. Chem. Soc. 78*, 751–755.

Porath, J. and Flodin, P. (1959). Gelfiltration: A method for desalting and group separation. *Nature 183*, 1657–1659.

Regnier, F. E. and Gooding, K. M. (1980). Review. High-performance liquid chromatography of proteins. *Anal. Biochem. 103*, 1–25.

Rosengren, Å., Bjellqvist, B., and Gasparic, V. (1976). A simple method of choosing optimum pH-conditions for electrophoresis. In: *Electrofocusing and Isotachophoresis, Proceedings of the International Symposium*, August 2–4, 1976, Hamburg, Germany. B. J. Radola, and D. Graesslin (Eds.), Walter de Gruyter, Berlin (1977), pp. 105–171.

Scott, R. P. W. and Kucera, P. (1979). Mode of operation and performance characteristics of microbore columns for use in liquid chromatography. *J. Chromatogr. 169*, 51–72.

Sjödahl, J. (1980). High resolution liquid chromatography of proteins. *Sci. Tools 27*, 54–57.

Sjödahl, J. and Winter, A. (1982). High speed gel filtration as a tool to optimize protein fragmentation. In: *Protides of the Biological Fluids. Proceedings of the Thirtieth Colloquium*, vol. 30, H. Peeters (Ed.), Pergamon Press, Oxford, pp. 693–96.

Sjödahl, J., Israelsson, R., and Ericsson, J. (1982a). An HPLC-pump optimized for the new areas of column liquid chromatography. *Sci. Tools 29*, 7–10.

Sjödahl, J., Lundin, H., Eriksson, R., and Ericson, J. (1982b). Solvent delivery in the new application areas of high performance liquid chromatography. *Chromatographia 16*, 325–329.

Tswett, M. (1903). O novoj kategorii adsorbeionnych javlenij i o primeneni ioh k biochimiceskomu analizu. *Proc. Warsaw Soc. Nat. Sci. Biol. Sect. 14*, No. 6.

Author Index

Arlinger, L., 179

Caron, M. G., 9
Chen, H.-C., 83
Cooper, B. F., 133
Crooke, S. T., 9

Dilley, K. J., 179

Eriksson, R., 179

Funae, Y., 59, 105

Greenhut, J., 133

Hara, I., 95

Ikemoto, F., 105
Iwao, H., 105

Kato, Y., 1, 149
Kim, S., 105
Kojima, K., 149, 157
Konishi, K., 43

Kotake, A. N., 59

Lefkowitz, R. J., 9

Nagatsu, T., 149, 157
Nakamura, N., 105

Okazaki, M., 95

Parvez, H., 1, 149, 179
Parvez, S., 1, 149, 179
Pellerin, J., 179

Rebar, R., 9
Rudolph, F. B., 133

Shimohigashi, Y., 83
Shorr, R. G. L., 9
Sjödahl, J., 179
Strohsacker, M. W., 9

Takagi, T., 27

Yamamoto, K., 59, 105

219

Subject Index

Ach R, 55, 56
Acid carboxypeptidase, 50
α_1-Acid glycoprotein, 35–38, 50
Adenosine deaminase, 142
Adenylate kinase, 34
β-Adrenergic receptor, 13–25
Aldolase, 55, 56
Anthracene, 213, 215
Aprotinin, 48
Aquapore CX-300, 135
Aromatic L-amino acid decarboxylase,
    151–154

BAM-12P, 159
BAM-20P, 159
BAM-22P, 159
Benzene, 213
Blue dextran, 48
μ-Bondapak C18, 158, 165, 167, 172
Bovine serum albumin, 33, 37, 44, 48, 55,
    56, 88, 112

Carbonic anhydrase, 37
Catecholamine-synthesizing enzyme,
    149–155
Chylomicron, 96–99
Chymotrypsinogen, 44, 112
CP-Spher C8, 165
Cyanogen bromide fragment, 45, 51
Cytochrome C, 44, 48, 112, 216
Cytochrome P-450, 59–81

DEAE-Glycophase G, 134
DEAE Sepharose 6B-CL, 13, 14
Denaturing solvent, 43–57
Dibutylphthalate, 205
Diethylphthalate, 205
Dihydropteridine reductase, 151–154
Dimethylphthalate, 205
Diphenyl-RP, 158
DNA fragment, 1–7
Dopamine-β-hydroxylase, 151

Dynorphin, 158, 159, 167
  –(1–8), 167
  –(1–13), 167
  –(1–17), 159, 165, 167
  –(1–24), 167
  B, 167

α-Endorphin, 158, 165
β-Endorphin, 158
γ-Endorphin, 158, 165
Enkephalin, 157–177
    containing peptide, 157–177
    precursor protein, 172
Enolase, 34
Ethyl-4-hydroxybenzoate, 205

Fetuin, 50
Fibrinogen, 55, 56, 206

γ-Globulin, 56, 206, 207
    H chain, 50
Glutamic dehydrogenase, 34
Glycopolypeptide, 49–50
Glycoprotein, 83–94
GPC/LALLS, 36–38
6 M Guanidine hydrochloride, 48–52, 88

HDL_2, 96–101
HDL_3, 96–101
HDL apoprotein, 102
Hemoglobin, 55, 56
Human cholrionic gonadotropin, hCG,
    84–93
Human serum, 191
    albumin, 190, 206
    lipoprotein, 95–103
Hypersil ODS, 167

Immunoglobulin G, 212, 216
Insulin
    A chain, 44, 48, 51
    B chain, 48

Lactic dehydrogenase, 34
β-Lactoglobulin, 44, 51, 55, 56
LDL, 96–101
Leu-enkephalin, 167, 172
LiChrosorb
    CN, 159
    Diol, 172
    RP-8, 159
    RP-18, 158, 165, 167, 172, 204, 205
β-Lipotropin, 158
Low-angle laser light scattering photometer,
    27–41
Lysozyme, 44, 51

Membrane-bound hormone receptor, 9–26
Membrane protein, 53
2-Mercaptoethanol, 48
Met-enkephalin, 158, 167, 172
    precursor, 158, 172
Methyl-4-hydroxybenzoate, 205
N-Methyltransferase, 151–154
Mitomycin, 212
Mono Q, 20–21, 134
Mono S, 134
Mouse serum, 200
Myoglobin, 44, 48, 56, 190, 206, 207, 216

Naphthalene, 213, 215
α-Neo-endorphin, 159, 167, 172
β-Neo-endorphin, 159, 165, 167
Nucleosil
    C18, 159, 165, 167
    Phenyl, 159, 167

Ovalbumin, 37, 44, 50, 51, 88, 112, 207,
    216
Ovoinhibitor, 50
Ovomucoid, 50

Partisil
    ODS, 158
    SCX, 158
Pepsin, 48, 51
Phenol, 204
Phenylalanine, 88
Phenylethanolamine, 151–154
Plasmid DNA pBR322, 1
Pre-prodynorphin (β-neo-endorphin/
    dynorphin precursor), 161
Pre-proenkephalin, 163
Pre-proopiomelanocortin (corticotropin-β-
    lipoprotein precursor), 160
Proenkephalin, 172
Pro-methionine-enkephalin, 158
Pro-opiocortin, 158

Propyl-4-hydroxybenzoate, 205
Protein column
    I-125, 10, 15, 16
    I-250, 10, 15, 16
PTH-amino acid, 210
Putative enkephalin precursor, 158

Renin, 105–131
    binding protein, 105–131
Ribonuclease, 37, 51
    B, 50
Rimorphin, 159, 167
RNA, 1–7

Salmonella phage P22 tail protein, 144
SDS, 45–48, 52–53
Sephadex
    G-50, 11, 12, 14, 16
    G-100, 86
Sepharose 4B-CL, 10
Spherisorb
    CN, 158
    ODS, 158
Substance P, 167
Sulfate binding protein, 143
SynChropak
    AX-300, 60–69, 135–145, 150
    CM-300, 135

Taka-amylase A, 50
Thyroglobulin, 56, 88, 112, 207, 216
Toluene, 213
Transferrin, 50, 56
Tripsinogen, 48, 51
Trypsin inhibitor, 88
TSKgel
    DEAE-5PW, 20–21, 107, 124–127, 134
    DEAE-3SW, 107, 122–127
    G2000SW, 2–7, 12, 23, 86–91, 150,
        172
    G3000SW, 2–7, 10–23, 32, 43–52,
        86–91, 98–102, 107–119, 127,
        150–153
    G4000SW, 2–7, 10–23, 43–53, 98,
        150–153
    G5000PW, 98–99, 107
    IEX535 CM, 107, 119–128
    LS-410 ODS SIL, 167
    PW, 96
    SP-5PW, 107, 121, 127, 134
    SW, 1–7, 96
Tyrosine, 190
    hydroxylase, 151–154

Ultrasphere

Octyl, 158, 167, 172
ODS, 158, 165, 167
Ultropac TSK
  545DEAE, 194
  645DEAE, 200
  2000SW, 201
  3000SW, 190, 206, 216
  4000SW, 191
  GSWP, 207, 212
8 M Urea, 51–52

Uremic peptide, 194

Vitamine B_{12}, 204, 207, 216
VLDL, 96–101

Zorbax
  CN, 165
  ODS, 165
  TMS, 159, 165